ΛD

Architectural Design
March/April 2009

Closing the Ga₁
Information Models in Contemporary Design Practice
Guest-edited by Richard Garber

IN THIS ISSUE
Main Section

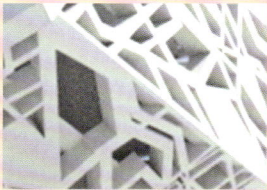

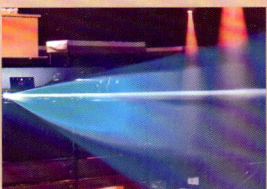
WILEY
wiley.com

CW01424511

Architectural Design

Vol 79 No 2
March/April 2009
ISBN 978-0470 998205

Editorial Offices
John Wiley & Sons
International House
Ealing Broadway Centre
London W5 5DB

T: +44 (0)20 8326 3800

Editor
Helen Castle

Regular columnists: Valentina Croci, David Littlefield, Jayne Merkel, Will McLean, Neil Spiller, Michael Weinstock and Ken Yeang

Freelance Managing Editor
Caroline Ellerby

Production Editor
Elizabeth Gongde

Design and Prepress
Artmedia Press, London

Printed in Italy by Conti Tipocolor

Sponsorship/advertising
Faith Pidduck/Wayne Frost
T: +44 (0)1243 770254
E: fpidduck@wiley.co.uk

Subscribe to AD

AD is published bimonthly and is available to purchase on both a subscription basis and as individual volumes at the following prices.

PRICES
Individual copies: £22.99/$45.00
Mailing fees may apply

ANNUAL SUBSCRIPTION RATES
Student: UK£70/US$110 print only
Individual: UK £110/US$170 print only
Institutional: UK£180/US$335 print or online
Institutional: UK£198/US$369 combined print and online

Subscription Offices UK
John Wiley & Sons Ltd
Journals Administration Department
1 Oldlands Way, Bognor Regis
West Sussex, PO22 9SA
T: +44 (0)1243 843272
F: +44 (0)1243 843232
E: cs-journals@wiley.co.uk

[ISSN: 0003-8504]

Prices are for six issues and include postage and handling charges. Periodicals postage paid at Jamaica, NY 11431. Air freight and mailing in the USA by Publications Expediting Services Inc, 200 Meacham Avenue, Elmont, NY 11003.
Individual rate subscriptions must be paid by personal cheque or credit card. Individual rate subscriptions may not be resold or used as library copies.

All prices are subject to change without notice.

Postmaster
Send address changes to 3 Publications Expediting Services, 200 Meacham Avenue, Elmont, NY 11003

RIGHTS AND PERMISSIONS
Requests to the Publisher should be addressed to:
Permissions Department
John Wiley & Sons Ltd
The Atrium
Southern Gate
Chichester
West Sussex PO19 8SQ
England

F: +44 (0)1243 770620
E: permreq@wiley.co.uk

CONTENTS

AD+

Editorial

Helen Castle

In the last couple of years, building information modelling (BIM) has been the subject of many professional guides and books.[1] This is a reflection of the highly practical concerns that face any firm considering the implementation of BIM processes. For the adoption of BIM – 'a single, intelligent, virtual model' that satisfies 'all aspects of the design process' (see Richard Garber's full definition in his introduction on p 8) requires a lot more than software training. It has far-reaching implications on construction operations and management, while its collaborative nature shifts conventional models of business, contractual liability and risk taking. So why should *AD* be concerned now with BIM?

BIM surpasses the nuts and bolts of practice. There is nothing half-hearted about building information modelling. BIM ushers in over-arching cultural and design changes: Dennis Shelden of Gehry Technologies describes it as a paradigm shift (p 80); Urs Gauchat regards it as a means by which architects can counter years of marginalisation and reclaim their rightful position near the top of the construction food chain (p 32); and Richard Garber likens it to a profound change analogous to that which architecture underwent in the Renaissance when master builders became architects (p 88).

So why should this group of software products bring changes previously not wrought in almost two decades of digitalisation? Coren Sharples, partner of SHoP Architects in New York, makes the enlightening comparison between the 'dumb' walk-through that limited clients to visualising the form of a building and BIM models that are embedded with constantly updated data that enable clients, contractors and consultants to be empowered with 'marketing, financial, operations and life-cycle costings information' (p 43). It is the very collaborative nature of the data feedback through the modelling process with its potential to inform and communicate that turns the tables on all existing systems.

All the contributors in this issue of *AD* are passionate advocators of BIM, but not reservedly so. Just as BIM has the potential to rein back architects' control over the design process, it could also dilute their influence through collaboration. Nat Oppenheimer of Robert Silman Associates structural engineers, a self-styled 'enthusiastic sceptic', for instance, fears that 'integration appears to be leading to oversimplified buildings' (p 101). For architects to truly take on the potential of BIM technologies, it has to be clear that they have to be prepared to truly work in new ways and restructure practice accordingly. This is exemplified by SHoP Architects who have used BIM not only to enhance collaborative processes with contractors and consultants and to keep their clients better informed with the type of data they require, but have also set up their own construction management firm, SHoP Construction Services, to further edge the design and delivery gap closer together. △D

Note
1. See, for instance: Dana K Smith and Michael Tardif, *Building Information Modeling: A Strategic Implementation Guide for Architects, Engineers, Constructors, and Real Estate Asset Managers*, ISBN: 978-0-470-25003-7, May 2009; Chuck Eastman, Paul Teicholz, Rafael Sacks and Kathleen Liston, *BIM Handbook: A Guide to Building Information Modeling for Owners, Managers, Designers, Engineers and Contractors*, ISBN: 978-0-470-18528-5, March 2008; Eddy Krygiel, Brad Nies and Steve McDowell (Foreword), *Green BIM: Successful Sustainable Design with Building Information Modeling*, ISBN: 978-0-470-23960-5, April 2008. All John Wiley & Sons Inc (Hoboken, NJ).

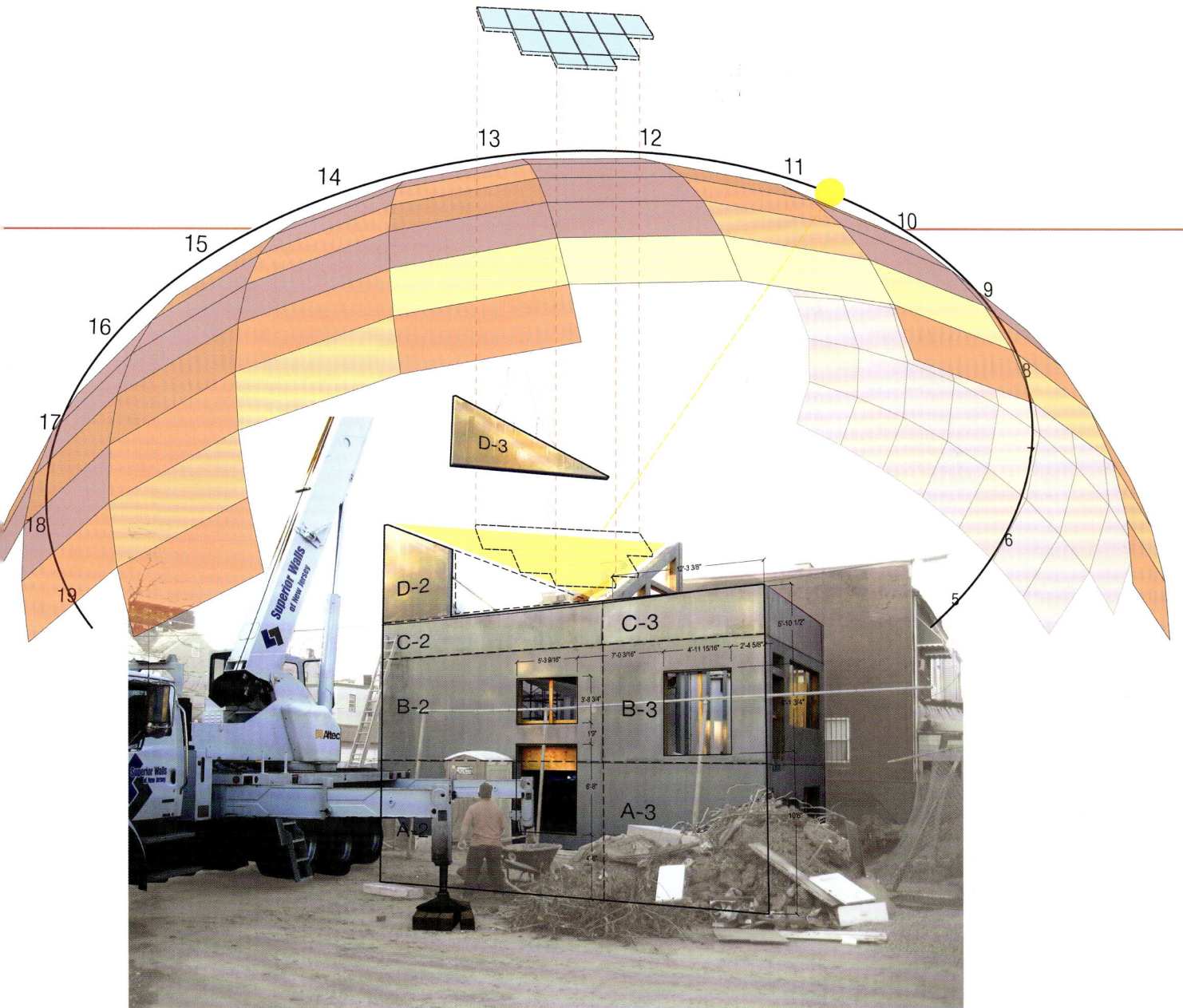

GRO Architects, PREFAB House, Jersey City, New Jersey, 2008–09

Precast-concrete wall components and the angle for roof pitches to utilise maximum solar collection were both studied using information modelling packages in a small, experimental house by GRO Architects. A digital model of the house was used to simulate the optimal orientation and location of photovoltaic panels and ensured that a continuous amount of electrical power would be self-generated, thus allowing the house to provide its own power while feeding unused current back to the city's electrical grid in certain months. The information model was also used to rationalise the precise shapes and joints of the concrete panels, here shown being craned into place. The panels were delivered to site in three truckloads and assembled via crane by a team of four workers over three days. The panels have metal studs cast into them for the fastening of interior surfacing and rigid insulation, which allows them to outperform conventionally constructed walls. The parametric capabilities of information modelling software allowed for the constraining of window and door openings, and maintained necessary dimensions from the edges of each panel. The multiple design and simulation checks performed here allowed the architects to engage earlier in the construction process and ensure more efficient construction schedules and a better-performing building.

Optimisation Stories

The Impact of Building Information Modelling on Contemporary Design Practice

By Richard Garber

Greg Lynn FORM, Garofalo Architects and McInturf Architects, Presbyterian Church of New York, Queens, New York, 1999
Greg Lynn's Presbyterian Church was 'stick-built'; that is, manually constructed of traditional construction materials that were fabricated and installed based on two-dimensional construction documents and shop drawings, much like a conventional project. However, it became a significant precedent for several of the architects represented in this issue of *AD* as the manual creation of construction documents in a CAD system required the measuring and rationalisation of the three-dimensional components of the virtual model. It was critical even at this early stage of the virtual construction to develop methods for the precise and accurate translation of three-dimensional geometries into a two-dimensional CAD document (profiles, plans, etc). The result is a stunning sanctuary space imagined in a virtual environment that, once actualised, spoke to the merit of such a design process while attracting the interest of a whole generation of young designers.

Over the last 25 years, the broad use of computer software has revolutionised the way we generate and document architectural design proposals. Much of this work has occurred either in visualisation or formal speculation (mainly academic concerns) or in conventional documentation and management (mainly professional preoccupations). This separation between theory and practice is amplified by the relationship between architects and those who build their design proposals. This formerly manual transfer of design documentation via the interpretation of others invites breaks, or gaps, in what should be a continuous and interrelated process of design development and building actualisation. However, these gaps are now closing as new possibilities emerging at either end of the spectrum demand the support of one another – advanced analysis software would not be necessary without design imagination.

Design computing has expanded in scope beyond representation-based documentation to now include analysis, simulation and digital fabrication. These new components allow architects to better understand and manage how their virtual ideas are realised, and to innovate or challenge traditional delivery and construction methods. The synthesis of such technologies and the need for better construction management has led to the emergence of building information models which close the gap and, in turn, promise to revolutionise contemporary design practice.

While current technologies are not sufficiently developed for full-scale buildings to be produced with computer numerically controlled (CNC) hardware, they do allow developed building information models to more precisely assist in the translation of a virtual construct into an actual one. Can a cost-effective paradigm shift be achieved using new computing technologies in architectural design?

The potential of building information modelling (BIM) is that a single, intelligent, virtual model can be used to satisfy all aspects of the design process including visualisation, checking for spatial conflict, automated parts and assembly production (CAM), construction sequencing, and materials research and testing. The model is shared, and contributed to, by all parties involved in the construction of the building, from architects to engineering consultants, contractors and subcontractors. This suggests a paradigm shift in design procedure and teaching that would involve time-based iteration and testing, not only of design potentials, but also of construction in a virtual environment. Indeed information models foster a kind of automatic coordination and collaboration that, partly

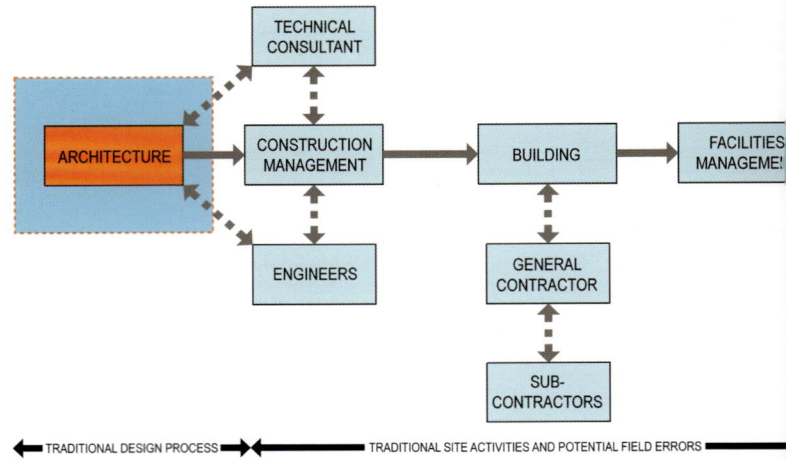

GRO Architects, Diagram showing the traditional architectural design process, 2008
In traditional architectural practice, a linear exchange of analogue information facilitated interpretation and the need for clarification on site. In this 'possible to real' paradigm, it was therefore necessary to transmit conventional documentation to others involved in the construction process. The manual translation of quantitative and qualitative building information (bills of materials, component quantities, as well as structural and environmental efficiencies based on rules of thumb) made full coordination of all the construction activities difficult. The result was a reality that was like the possible only through representation.

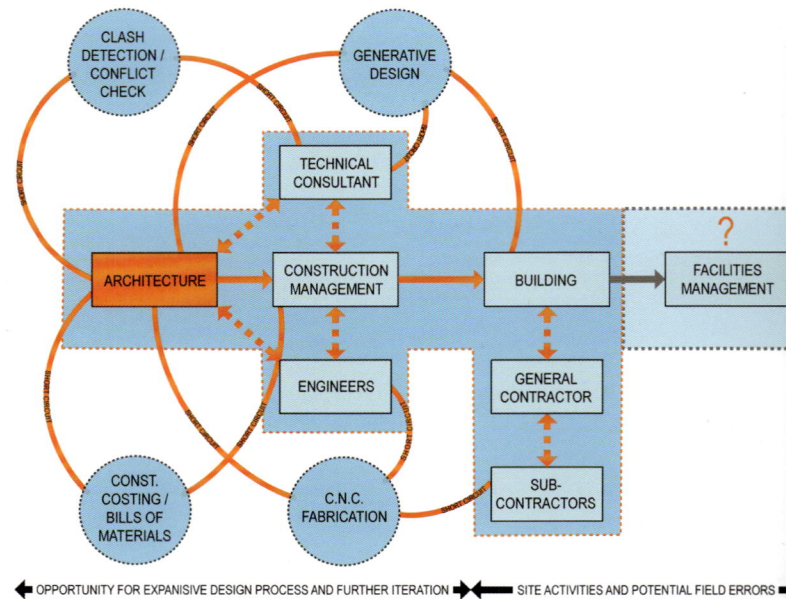

GRO Architects, Information modelling in architectural practice, 2008
In architectural practice augmented by information modelling, virtual translation of data and real-time performance testing becomes possible. In this new paradigm – the 'virtual to actual' – the virtual is already fully real and needs only be actualised by a process of translation from virtual matter to physical matter. Efficiencies of computer numerically controlled (CNC) fabrication, real-time generation of bills of materials and schedules, as well as construction sequencing occur. Through information modelling, the design process becomes augmented and more iterative, and allows for more in-depth collaboration between architects and allied design professionals such as engineers, technology consultants and subcontractors in a non-linear fashion. Streamlined management of construction projects is not the only impact of these technologies as many information modelling packages contain generative design modules that promise further formal speculation tied to performance. Information models also lend themselves to the life-cycle management of a building, well beyond the conclusion of the design and construction process.

due to the medium and partly due to the intentions of designers, has not been seen before in the broader building industry. However, this seems less likely to specifically enhance the architect's position as a central hub through which all things pass – a return to the status of master builder – during the design of buildings. Perhaps, more interestingly, these technologies allow for a medium in which notions of creativity and innovation merge through performance operations, cost efficiencies, and material and system simulations that are iterated digitally throughout the design process as opposed to in the field during construction, where finding errors or conditions not properly coordinated could lead to costly expenditures in terms of time and budget.

Information Models in Contemporary Design Practice

According to the US National Institute of Building Sciences (NIBS), BIM refers to 'the use of the concepts and practices of open and interoperable information exchanges, emerging technologies, new business structures and influencing the re-engineering of processes in ways that dramatically reduce multiple forms of waste in the building industry'.[1] The institute's work focuses mainly on the further development of integrated product delivery (IPD) and data translation for the streamlined sharing of information between architects and subcontractors, such as structural steel fabricators. Other areas in which it is developing standards include automated code compliance checking (AC3) and construction to operations building information exchange (COBIE).

Many firms are finding attractive information models based on their digital management capacity. As digital databases, these models can control and monitor construction events so as to streamline and refine the construction process itself. There are already numerous examples of how BIM has saved time and made the building process more efficient: one only needs to be reminded of how the geometrically complex Denver Art Museum by Daniel Libeskind was completed in 2005 by the Midwestern US contractor MA Mortensen three months ahead of schedule, and with no change orders during the construction process.[2]

The ability of an architect, engineer or contractor to simulate construction in a digital model has many merits and uses 4-D, or time-based, operations in what is seen as, perhaps finally, the use of animation for something other than the formal manipulations of the 1990s. Originally conceived of by Greg Lynn and others[3] for its capacity in the design process to allow all geometry to respond to a series of programmable

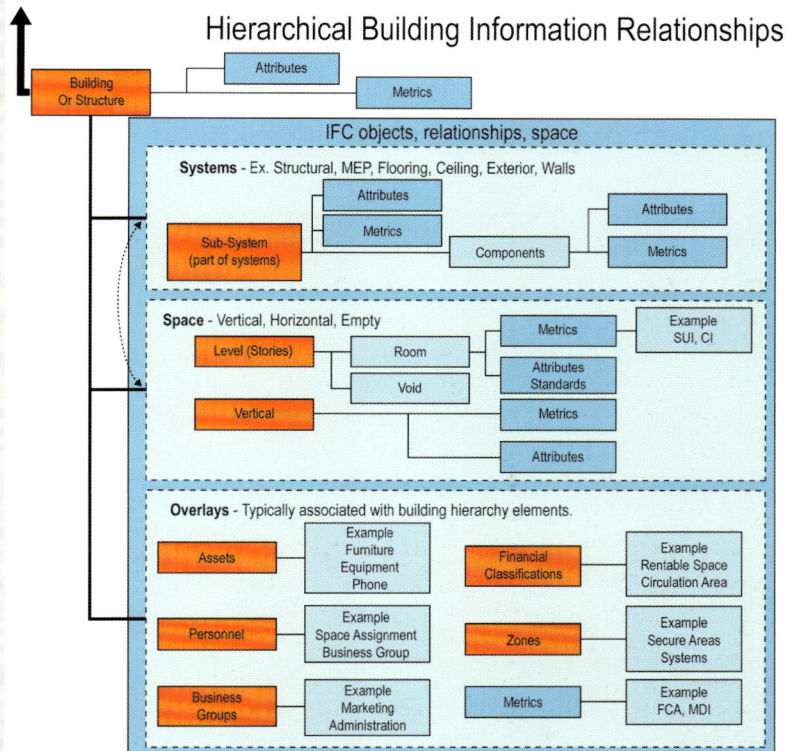

Hierarchical Building Information Relationships

National Institute of Building Sciences, Hierarchical building information relationships, 2006
The National Institute of Building Sciences, based in Washington DC, has been instrumental in setting building information modelling (BIM) standards in the US and internationally. It has divided the BIM process into an object-based system of Industry Foundation Classes (IFCs): Systems represent the physical entities of the building and can be classified by form (high rise, suspension bridge) or function (single-family residence, courthouse); Space is tied to the physical structure or systems by a boundary definition; and Overlays are the abstract data – organisational, operational, functional, financial, non-fixed assets, resources, personnel, etc – tied to the Systems and Space classifications. Further definitions can be found on the institute's website (www.nibs.org).

Precast insulated concrete panel
(PIPs)

Cedar rain screen

Low-emittance glazing

260 SF Photovoltaic Panels oriented
due south

300 SF modular green roof beyond
allowable building set-back

Modular furniture

Radiant heating in all concrete
slabs

Precast insulated concrete panel
(PIPs)

GRO Architects, PREFAB House, Jersey City, New Jersey, 2008
BIM systems have two important and interrelated aspects. First, they allow all virtual geometry to be linked to real-time databases for the accurate costing of materials and for ensuring building components are properly integrated. Next, they allow for the smooth transfer of data to the software packages that enable simulation to occur. The PREFAB house was conceived as an economical yet environmentally responsive structure able to generate its own power and utilise efficient materials. In addition to developing a set of construction documents in a BIM system, the information model was translated into environmental analysis software to study criteria such as solar gain and heat loss in winter months, and also into a proprietary software to generate the 3-D formwork for the house's precast-concrete walls. The proprietary software also allows the virtual sequencing of the panel assembly prior to components being delivered to the site.

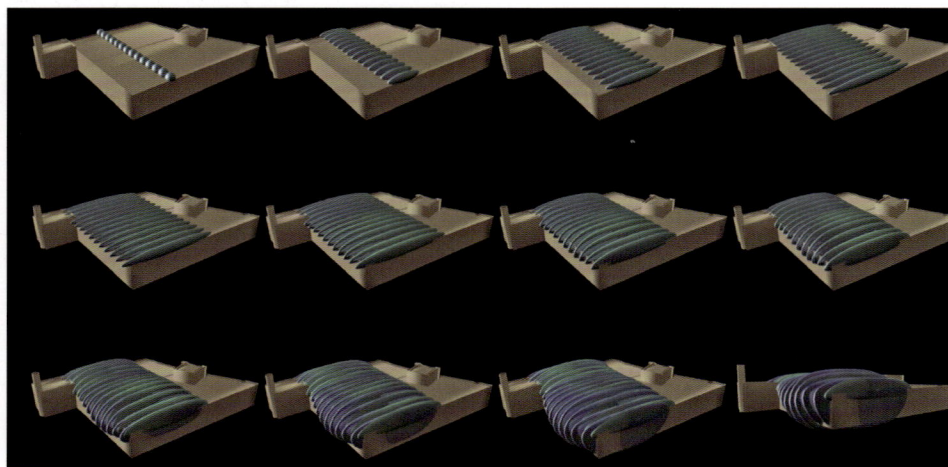

Greg Lynn FORM, Garofalo Architects and McInturf Architects, Presbyterian Church of New York, Queens, New York, 1999
The use of animation in early yet sophisticated modelling programs enabled designers to employ time-based operations in the design decision-making process. While these early modelling programs did not have the capability to link information such as material resistance to this form generation, virtual information such as mass and thickness could be extracted directly for material consideration. For the Presbyterian Church of New York, Greg Lynn FORM generated a virtual model that animated an interrelated series of forms for the main sanctuary space of the building.

(typically numeric) factors, animation became synonymous with the 'stopping problem'. This term was generally attributed to time-based geometric experiments in which the resulting forms were somehow imprecise because it was difficult to conventionally ascertain why one geometric frame was any more applicable than another: an arbitrariness that quickly branded certain architects and designers as not being grounded in the true problems of design and construction. However, these models were precise in that they were measurable (mass, volume, dimension, curvature) virtual constructs and represented a range of options. This notion of range forms the essence of the parametric experiments possible with BIM systems today: through optimisations of shape, cost, material, orientation and so on, an informed choice can be made from a family of related virtual solutions, and also serves to de-emphasise what perhaps was initially an interest in the formal aesthetic that also emerged from early studies of form generation and animation

Material Simulation and Testing

Still other capacities of BIM systems, such as clash detection, the generation of bills of materials and real-time construction schedules, and the transfer of digital data to fabricators or subcontractors, are attractive in their potential to optimise the design and building process. If construction can now be controlled through a single model or, more appropriately, a database that can accept information from a variety of sources, design optimisations such as those listed above can be studied in the earlier stages of development. The ability to respond to material or geometric factors such as stress, weight, hardness, volume and area, as well as time-based concerns such as sequence, has allowed designers to apply virtual attributes that yield form. Buildings can be understood according to how they perform as opposed to what they look like.

While there is still work to be done on the translation of various types of data into such a model, it is important to recognise that not all information needs to be distributed to the entire construction team at any given time: for example, a CNC fabricator responsible, say, for custom steel connection plates need not be concerned with the size and orientation of supply ducts as long as he or she can be 'digitally assured' that the ducts will not interfere spatially with the plates.

External Pressures

One of the more curious occurrences in the last two decades has been the emergence of new building 'specialists' whose purpose it is to oversee the complex and messy construction process. The impact of these construction managers seems unclear and is in some ways ironic. They are essentially watchdogs employed by owners or contractors to ensure economic and production transparency, yet often make the building management process more complex while also diverting fees away from those involved in the design of a building. The promotion of a more transparent process of checks and balances through information models would thus seem to be an important value-adding capacity of BIM.

The time-based, parametric and generational capacities of many information modelling softwares allow architects and designers to challenge conventional and outmoded construction methods while simultaneously introducing new techniques for organising and creating form and space. Parametric and organisational scripts for form generation take into account material attributes such as weight and maximum curvature so that the tedious process of translating one's design intentions into something buildable can become much more refined. This also offers the potential for the rationalisation of new building forms in much the same way that BIM packages can be used to optimise more conventional typologies or construction methods. Though the generational capacities of these tools are often overlooked, they nevertheless represent a real way of introducing the new organisations necessary to contend with increasingly complex, mixed and various programmes and building types to our constructed landscape.[4]

GRO Architects, Best Pedestrian Route, Lower Manhattan, New York, 2007
As BIM systems allow designers to think through problems of construction sequence and assembly virtually, the relationship of design to fabrication becomes much more comprehensive. CNC fabrication from digital files enables different parts to be cut from materials as economically as standardised parts were machined in the 20th century. In the fabrication and assembly of Best Pedestrian Route, a digitally fabricated temporary construction walkway, a series of uniquely shaped aluminium gussets were required to resist the bending moment of the cantilevered roof of the walkway. The gussets were water-jet cut from pieces of 6-millimetre (0.25-inch) and 12.5-millimetre (0.5-inch) aluminium sheets and precisely positioned within pockets that were CNC-routed to the inside faces of the plywood rib geometry. The precision with which the rib assemblies fit together could not have been achieved manually.

Interpretation, or at least discussion, on the project site of the Pavilion Zürichhorn, Switzerland, in 1960

In the possible to real paradigm, interpretation is necessary to construct an actual building from a set of design possibilities. As there is a gap in the transfer of static information between the architect and the general contractor, it is incredibly difficult to ensure the contractor has a comprehensive understanding of the design intent, and the architect of the construction techniques and methods. Interpretation leads to errors in the field, or situations where certain components or construction sequences need rethinking, both of which can be costly. This, unfortunately, has happened to the best of us! Pictured are Willy Boesiger, Editor of *Oeuvre complète*, City Architect Adolf Wasserfallen, Le Corbusier, Inspector of Public Parks Pierre Zbinden, and the gallery owner Heidi Weber.

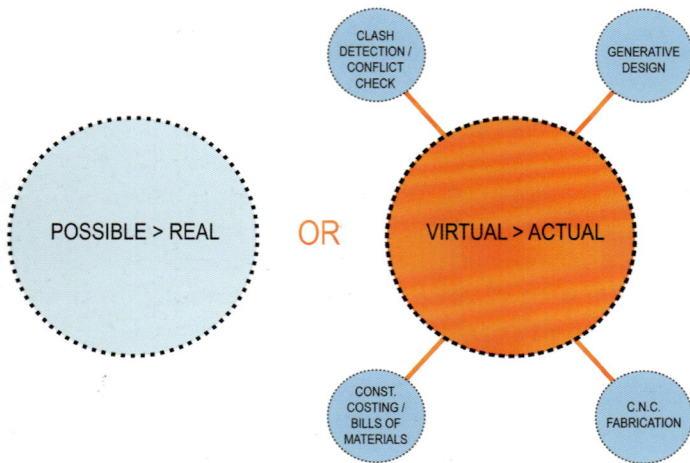

GRO Architects, Diagram showing the virtual to actual paradigm, 2008

In the possible to real paradigm, representations of a possible object or building are produced and transferred to another party for interpretation; because of this disjunction there is no way of ensuring that the possible and real will be the same. This traditional method has largely been two-dimensional; even with digital design techniques, the computer is used as a representational tool in the generation of renderings (pictures) or 2-D drawings of a building. This method separates the designer from the fabrication or construction process, and can be traced back to Leon Battista Alberti's decree that there should be a separation between the practices of design and making. As such its evolution has largely taken place in architectural schools. In contrast, the virtual to actual paradigm is an emerging method where (digital) material resistances can be tested through simulation. It is based on the translation and/or dissemination of a virtual 3-D construct, an example of which is translation to a CNC machine for direct manufacturing.

Douglas Gauthier and Jeremy Edmiston, BURST* Project LLC, BURST*008, 'Home Delivery: Fabricating the Modern Dwelling', Museum of Modern Art, New York, 2008

Delivery and erection of the *BURST House's CNC-cut structurally insulated panels (SIPs) to the exhibition site where they were fastened together. This process is a different take on prefabrication, in that the house arrives on site as a series of parts that are put together like a jigsaw puzzle. It is important to note the difference between this process, in which many unique parts are cut by a CNC machine, and one of mass standardisation where the same parts are assembled by workers on site as a series of repetitive tasks. BURST*008 used CNC output for efficient mass-customisation, demonstrating how information models can be employed to optimise, simulate and make construction methods more efficient.

The Virtual to Actual Paradigm

Though instrumental in closing the gap between design and building, preliminary, or schematic, opportunities are often overlooked in the adoption of parametric building information models in professional practice. Design theorists such as Sanford Kwinter and Manuel DeLanda have both highlighted the paradigm shift, made possible by information models, whereby the previous method of architectural delivery, the 'possible to real', is being supplanted by a new and seamless one, the 'virtual to actual'. In the first, the formulation of the initial design intention was a necessarily cerebral operation.[5] The designer would dream up a form and then represent it as an architectural proposal to which would be attached information about materials, construction methods and so on. Because proposals were generally documented as two-dimensional abstractions of a possible building, they required interpretation by others than the designer to realise the building. As such it was impossible to guarantee that the intentions of the architect would be precise in the built project.

An example of this conventional model in digital architectural design is image mapping, in which representational as opposed to performance-based criteria are used in the selection of an appropriate design proposal. Three-dimensional modelling programs were first used to simulate what a building would look like via libraries of 'materials' (generally bitmapped images) that could be applied with relative ease to preconceived forms. These forms were primarily variations of non-eidetic geometries (planes, spheres, cubes) that had no relation or resistance to the representations of materials they would receive. The outcome of such operations was largely deemed successful if the applied image maps were adequate in texture and scale. Whether brick was a suitable material choice for such a form, or masonry construction appropriate, was usually given far less consideration.

This virtual reality further separated architects from the material process of building. When not linked to a database or library, the bitmapped brick (or grass, granite, or 'purple', for that matter) wall has no properties or attributes that architects must consider in its construction. It has no weight, is unaffected by its height to thickness, and displays no resistance if a second virtual object is placed on top of it. Consequently, it has no reference or attribute data to determine whether or not it can sufficiently behave like a wall.

An often overlooked aspect of the introduction of computing to the design and management of buildings is the necessary break from traditional ideas and methods of teaching architecture. Though we are now some two decades into the digital, one senses that yet another gap between the teaching of new modes of practice using information models and traditional architectural education has been created. Lack of buildability and formal whimsy seem to inform the critique articulated by those who advocate more traditional methods of education. It is important to note, however, that the formal speculations of the last decade partially laid the foundations for the development and adoption of BIM systems in architectural design. When measured against performance or other optimising criteria, these speculations seem to intersect with the more pragmatic management tools and construction simulations proffered by information models.

In the contemporary virtual to actual paradigm, interpretation is no longer required because digital information models are already inherently real. As precise three-dimensional concepts are designed, tested, iterated and optimised in virtual space, they need only to be translated, or actualised, into physical media. A simple example would be a series of panels rationalised on a virtual sheet of plywood to be CNC-cut by a router.

Closing the Gap

The architects, engineers and academics who have contributed to this issue of *AD* have a broad and varied interest in the use of BIM technologies. The scope of the issue therefore covers architectural education, comparisons with historic models of practice, changes in contemporary design practice and the still newer opportunities for optimisation and CNC output. Rather than concentrating on a narrow or specific definition of BIM, the intention is instead to demonstrate how widespread and significant the impact of these technologies on the practice of architecture will be. As clever designers use BIM tools to better optimise, iterate and test their ideas, we will start to see design proposals and built projects that are not only integrated and precise, but truly new. ∆

Notes
1. See http://www.nibs.org/newsstory3.html/.
2. I described this process in a lecture at the School of Architecture at NJIT on 19 September 2005.
3. See Greg Lynn, *Animate Form*, Princeton Architectural Press (New York), 1999.
4. See Chuck Eastman, 'Automated Assessment of Early Concept Designs', pp 52–7 of this issue.
5. See Manuel DeLanda, 'Philosophies of Design: The Case for Modeling Software', in *VERB Processing: Architecture Boogazine*, Actar (Barcelona), 2002, pp 131–43.

Best Pedestrian Route

Lower Manhattan, New York, 2007

GRO Architects

By Richard Garber and Nicole Robertson

The process of construction is a liminal activity. This can be understood within the physical space and time between construction zone and building, the interface between digital techniques of fabrication and interpretation via contracting practices, or simply between the graphic symbol and diagrammatic pattern. Best Pedestrian Route (BPR) occupies and resonates within this threshold through multiple scales and meanings: a slanted and skewed temporary walkway, its cantilevered form is not only expressive of the volatility of its location, but is also, by design, able to systemically respond to the inherent instability created by the ongoing construction in Lower Manhattan.

Inspired by Jersey barriers in the US, which temporarily establish the boundaries of a road or construction site, and the graphic symbols of construction zones including two-dimensional diagonal stripes, arrows and signage, BPR is one of the most ambitious projects selected as part of the re:Construction Pilot Program sponsored by the Alliance for Downtown New York (ADNY) and the Lower Manhattan Cultural Council (LMCC). As the competition called for both a proposal and the selection of a possible site for the project, it was necessary to conceive of BPR as a virtual assembly that established diagrammatic relationships between its parts, external conditions, sidewalk obstructions such as street signs and phone booths, and the changing nature of construction sites themselves. However, as the final location was to be determined after the winning entry had been decided, it was critical that these relationships were not fixed, but able to respond parametrically as a system to a series of potential locations. The component parts were all digitally prefabricated on a computer numerically controlled (CNC) three-axis router, making assembly of the temporary structure possible in the course of three days.

BPR is conceived as a system of walkways, this being the first of several variations planned throughout Lower Manhattan, allowing visitors to shift their attention from the disruption of construction to a projected future downtown. ∆D

opposite: Inspired by the graphic symbols of construction sites and the instability of spatial and temporal boundaries of these zones, this temporary pedestrian walkway blends in with its surroundings. The appearance of instability is produced by its cantilevered and slanted form, which challenges visitors to question the definitions of construction and architecture, process and completion, temporary and permanent. As a prototype for a new kind of construction barrier, information modelling logistics are tested through the flexibility embedded in BPR's modular and parametric design, allowing this iteration to be the first of those planned.

right: Tool-paths are generated virtually and transferred to a CNC machine for cutting or forming. They are contained in ASCII text files known as G or M codes. G-Code is a functioning code that lists the tasks to perform the actual cutting or forming of materials, while M-Code, or machining code, tells the machine to start, stop, provide coolant and so on. Shown here is a sample of the code generated to cut the pictured rib pieces. The function S sets the spindle speed of the cutting tool, G0 directs the tool to rapidly move to the first position of the path, and G1 is linear interpolation, in which the machine finds an X–Y axis of the curves it is about to cut. For BPR, three methods of cutting were used: first, subtractive holing, or drilling (G81) to bore holes for bolting; next pocket cutting to accept aluminium gussets at areas of high bending moment; and finally profiling to precisely cut the final shape of ribs (both G90, G17, G40 functions). The X, Y and Z coordinates plot the path of each part.

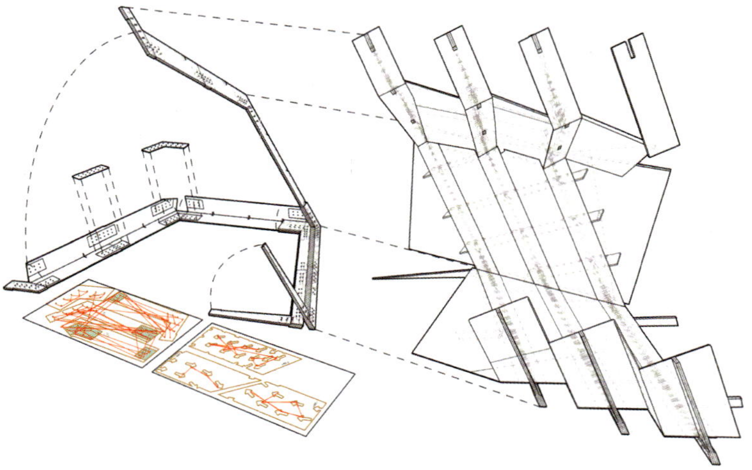

The fabrication and assembly sequence of Best Pedestrian Route began with the CNC-milling of marine-grade plywood parts in the FABLAB of the School of Architecture at the New Jersey Institute of Technology. Three specific types of tool-paths were utilised. First, subtractive holing was coded with a 5/16-inch diameter drill bit to create holes where the various plywood structural parts that form the ribs were to be bolted together. The 20-millimetre (0.75-inch) plywood sheets were bolted in thicknesses of 37.5 millimetres/1.5 inches (two sheets) or 75 millimetres/3 inches (four sheets) depending on moment stress along the length of the ribs themselves. To further strengthen these assemblies, pocket cuts removed material from the rib components for the addition of aluminium gusset plates sandwiched between sheets. All parts, including the gussets, sheathing and cladding for BPR, were then profile cut from 4 x 8 plywood sheets to their final, optimised shapes for assembly.

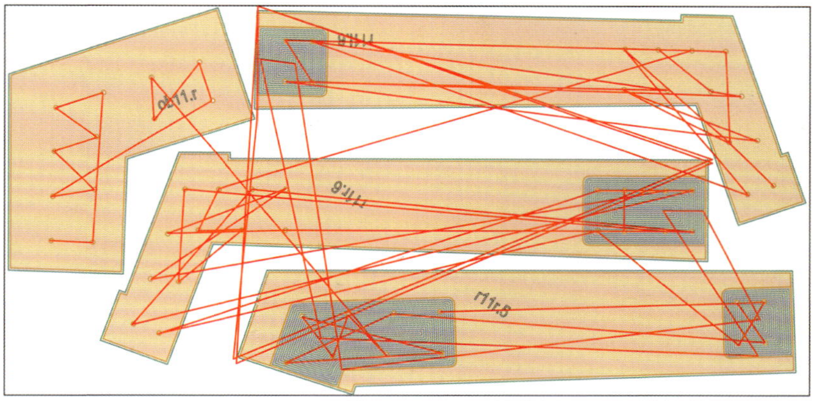

```
% S19000 G90 G17 G40 M3
(2 1/2 Axis Pocketing) 8th
S10000M3
G0 Z1.
X78.623 Y22.685
G1 Z1.025 F400
Z1.
X79.448 Y22.684 Z0.7 X79.447 Y22.487 X73.101 Y22.501 Y22.693 X79.448 Y22.684 X79.573 Y22.809
X72.976 Y22.819 Y22.376 X79.572 Y22.361 X79.573 Y22.684 X79.698 Y22.683 X79.697 Y22.236 X72.85
Y22.251 X72.852 Y22.944 X79.698 Y22.934 Y22.683 X79.823 Y23.058 X72.727 Y23.069 X72.725 Y22.127
X79.822 Y22.111 X79.823 Y22.683 X79.948 X79.946 Y21.986 X72.6 Y22.002 X72.602 Y23.194 X79.949
Y23.183 X79.948 Y22.683 X80.073 X80.074 Y23.308 X72.477 Y23.319 X72.475 Y21.877 X80.071 Y21.86
X80.073 Y22.68 X80.198 X80.196 Y21.735 X72.349 Y21.75 X72.353 Y23.445 X80.199 Y23.433 X80.198
Y22.683 X80.323 Y22.682 X80.324 Y23.558 X72.228 Y23.57 X72.224 Y21.628 X80.321 Y21.61 X80.323
Y22.682 X80.448 X80.445 Y21.485 X72.099 Y21.503 X72.103 Y23.695 X80.449 Y23.682 X80.448 Y22.682
X80.573 X80.575 Y23.807 X71.978 Y23.82 X71.974 Y21.378 X80.57 Y21.359 X80.573 Y22.682 X80.698
X80.695 Y21.234 X71.849 Y21.254 X71.854 Y23.945 X80.7 Y23.932 X80.698 Y22.682 X80.823 Y22.681
X80.825 Y24.057 X71.729 Y24.07 X71.723 Y21.129 X80.82 Y21.109 X80.823 Y22.681 X80.948 X80.944
Y20.983 X71.598 Y21.004 X71.604 Y24.196 X80.95 Y24.182 X80.948 Y22.681 X81.073 X81.076 Y24.307
```

Illuminated at night, the cladding of this temporary pedestrian walkway emits light through arrow-shaped apertures that recall the graphic symbols of construction signs while maintaining the structural integrity of the panels by acting as a diaphragm between ribs. The dynamic flow pattern of the arrows plays on the experience of disorientation created by the extensive construction zone in Lower Manhattan. This zone encompasses more than 80 individual sites and work is expected to be in progress until at least 2012, giving the usually temporary environment of construction a permanence that requires thoughtful design consideration.

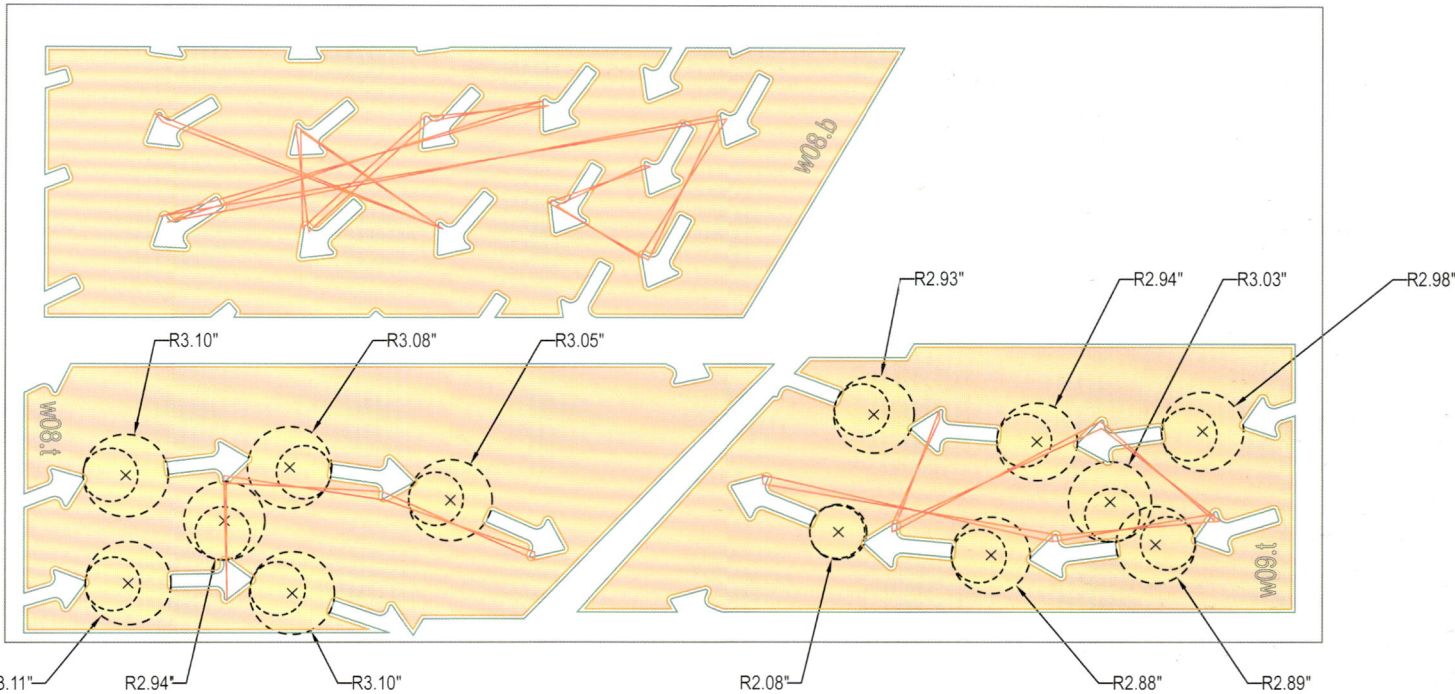

The illuminated school of arrows which flow across the cladding panels was generated by plotting a scripted series of points across the length of the iconic orange and white sheathing. BPR was designed with a roof cantilever ranging from 1.5 to 3.3 metres (5 to 11 feet), and the sheathing between the cantilevered ribs in the walkway's roof needed to resist lateral wind loading and prevent sideways movement of the cantilever. To achieve this while maintaining the school of arrows, individual cut-outs were constrained so that they would be separated by a minimum radius of 5 centimetres (2 inches), thus maintaining lateral stability in each roof panel.

The eight marine-grade plywood ribs that form the structure and geometry of Best Pedestrian Route were assembled in the construction staging area of the future Fulton Street Transit Hub adjacent to the Corbin Building in Lower Manhattan. Here, New Jersey Institute of Technology (NJIT) students Justin Foster and Ninett Moussa work with NJIT assistant professor and founder of GRO Architects Richard Garber to tighten the bolts that secure the multiple layers of plywood, glue and pocketed aluminium gussets.

BURST*008
Museum of Modern Art

New York, 2008

Douglas Gauthier and Jeremy Edmiston

By Douglas Gauthier

SIP Exterior Skin

Plywood Interior Skin

Glazing

Facade, Clerestory & Windows

Vents & Benches

SIP Bowtie Walls

Decking & SIP Floors

Steel Clips

Plywood Ribs

Moment Frames & Foundation

The Grid

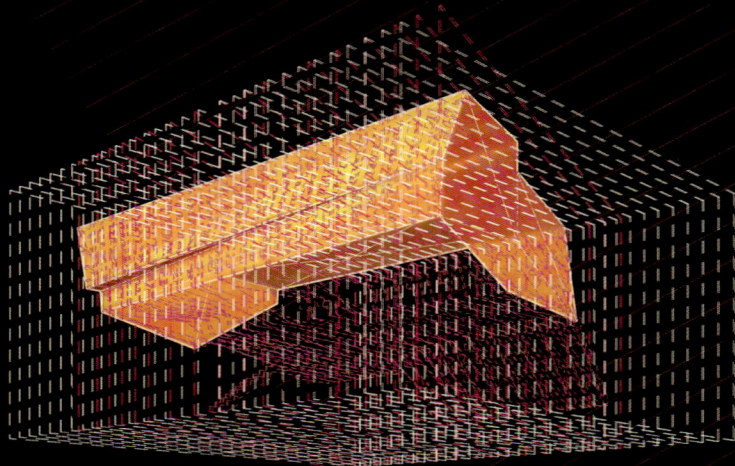

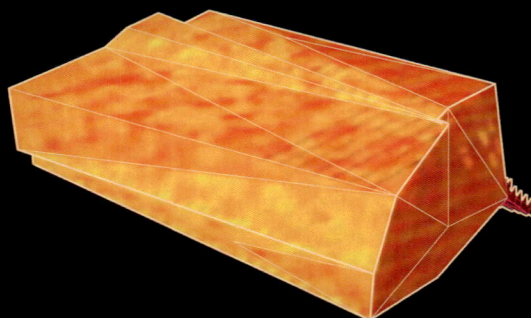

BURST* is a prefabricated housing system that functions like a kit of parts to produce homes that use small building pieces to achieve individually tailored spaces, allowing the architectural shape of each house to conform to the specifics of distinct project constraints. An alternative to mass-produced versions of domestic life that reduce prefab houses to differing arrangements of boxes, each BURST* system offers the potential of creating unique spaces and forms based on environment, site, orientation and the needs of the owners. This is achieved using generative technologies to expand the range of architectural forms for domestic and inexpensive construction.

The structure and environmental mechanics of the BURST* houses were designed in collaboration with the New York office of Buro Happold Consulting Engineers. BURST* is an adaptable design process which responds to various sites, climates, owners and programmes. The flexibility of BURST* comes from literally weaving two sections together – the natural ground plane and an artificial, manipulated plane. The two planes travel vertically and horizontally to comprise the ground, the floor and the walls. The system uses only passive means to maintain temperature comfort levels; depending on the specific conditions for an individual house, the weave can open, close and reshape in order to allow in or keep out warming sun and cooling breezes. This environmental weave is reflected in the tension structure of the plywood ribs and structural insulated panel (SIP) skin. Similar to a kite or an aeroplane wing, the tension structure's high strength and lightness combine precisely for a project with a less than 5 per cent waste factor. The bulk of the construction process is achieved digitally, during which the geometry of the house and the individual pieces – structural ribs, walls, floors – are resolved, then precisely cut and numbered before BURST*ING on to the site.

BURST*008 was unveiled at the 'Home Delivery: Fabricating the Modern Dwelling' exhibition at the Museum of Modern Art (MoMA), 20 July to 20 October 2008. The exploded axonometric drawing on the right illustrates the massing, structure, and skin surfaces that, through geometric processes, converge to form the ribs from the geometry of the massing. The exploded axonometric drawing on the left illustrates the construction elements of the building. These elements include the structural grid, foundation and moment frames, plywood ribs, the steel X-clips, various structural insulated panels (SIPs), glazing and steel elements, as well as the interior and exterior skins. Because the house is made from relatively small elements, it can be responsive to structural, programmatic and environmental parameters: if the programme changes, the massing changes by shrinking or swelling; if the structural requirements change, the weave tightens and loosens; and if the location or environment changes, the apertures and openings of the house contract or dilate as required. The drawing was on display inside the house during the exhibition.

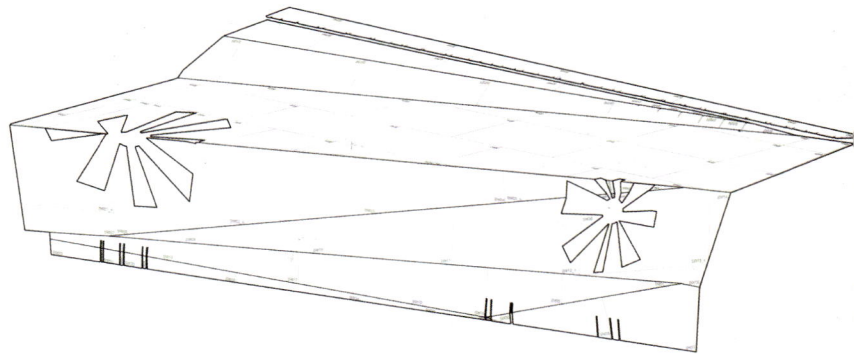

Exploded axonometric drawing – part of the assembly documentation for BURST*008. The drawing serves as a reference for locating and installing the interior wall and ceiling panels in the sleeping quarters. The flower-shaped apertures in the skin combine to form windows on the street, or north, side of the house. The MoMA curators questioned what they considered the Venturi-esque surface condition of the initial BURST* graphic, and were pleased and surprised when as a result it evolved into functional elements.

BURST*008 was fabricated as three structural rib assemblies that were trucked to, and then joined together on, the MoMA site. Here the second structural rib assembly is expanded and located on the moment frames via a crane.

Ribs nestling into the SIPS, illustrating the precision cutting and connective knot connections of the BURST* house. It is this weaving together of small parts that allows the house to adapt so readily to various changing criteria. Through small shifts in the design process, connection points can be moved to accept more weight, to allow a larger space, to open to the outdoors. The design method is then able to consider the overall sphere of domesticity as a spatial negotiation between programme, structure, natural and synthetic environments, time and materials.

The undercroft area of the house provides an entirely functional secondary space that may be used, for example, as a playroom, as an extra storage area, or as a welcoming area. In occupying the plot, the house is aware in its orientation, considering sun, wind and humidity in order that these conditions add to its productivity rather than remaining insignificant.

The north facade of the completed building. ∆

Text © 2009 John Wiley & Sons Ltd. Images: pp 18-19, 21 © BURST* Project LLC; p 20 © Anthony Rosello – Certain Pictures

The Future of Information Modelling and the End of Theory

Less is Limited, More is Different

The computational design strategist **Cynthia Ottchen**, who was previously Head of Research and Innovation at OMA, offers insights into the future of building information modelling (BIM). Now in the Petabyte Age of the data deluge, she argues that in our adoption of BIM we have to surpass mere data collection and technical optimisation and open up new ways of thinking with the creative use of 'soft data'.

OMA Research and Innovation Parametrics Cell, Study for Phototropic Tower, 2008
This digital rendering of the Phototropic Tower (Iteration_01) collaged into its Manhattan context shows the effect of a building volume and form based on a strategy of privileging and scripting selected data (views out of the building, increased light, minimal heat gain) over normative constraints. In this case, the resulting design contrasted with its context in terms of shape, setbacks and density, and did not stay within the typical constraints of economic, cultural or prescribed municipal rules including maximum building volume or floor area ratio, self-similar urban patterns (continuous street frontage) and zoning envelopes.

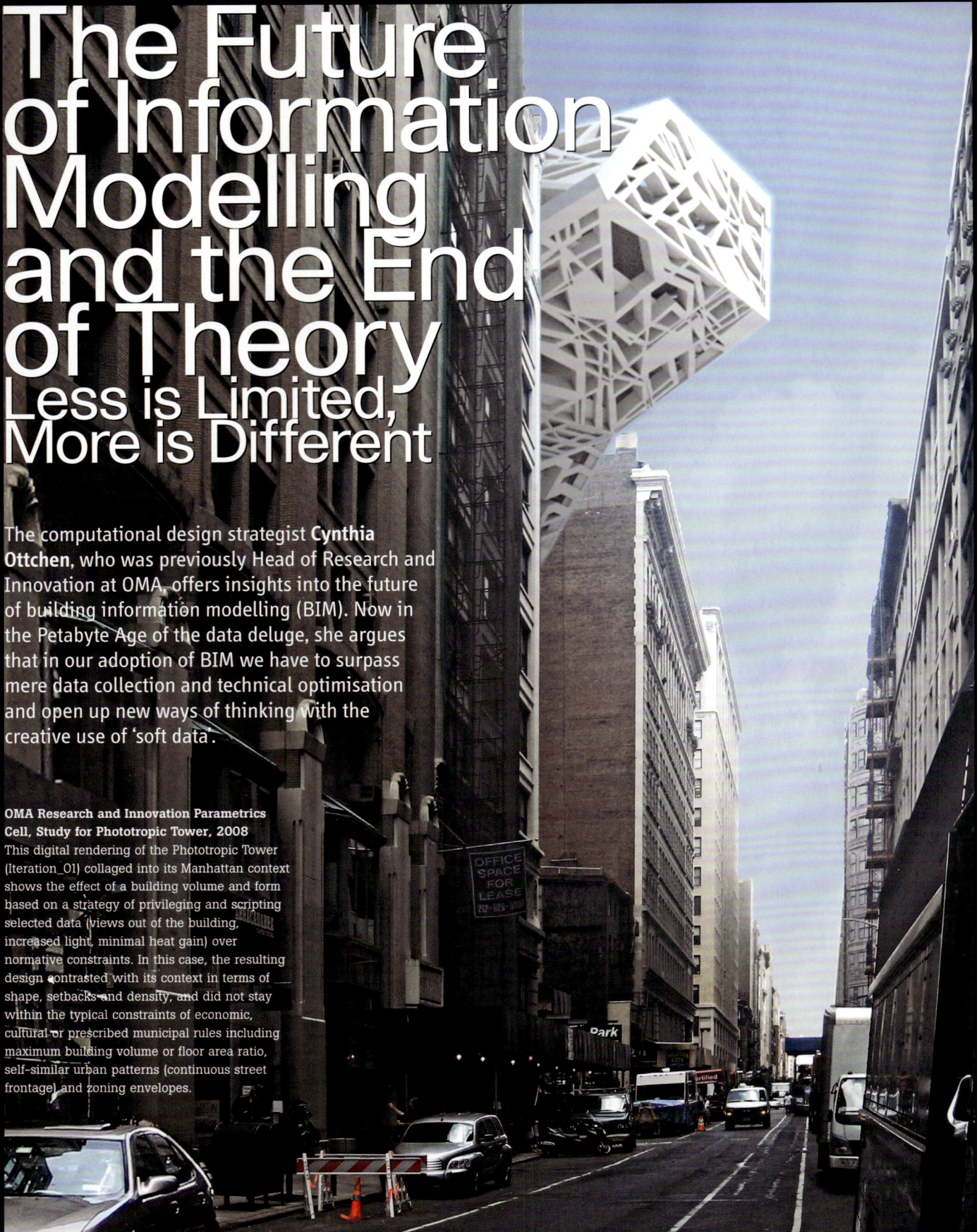

All models are wrong, but some are useful.
George Box, *Empirical Model-Building and Response Surfaces*, 1987[1]

The exploration of digitally generated architectural forms began in the architectural schools as early as the 1980s with the appropriation of animation software from the entertainment world. Such software lent itself to experimentation with notions such as fluidity and dynamics. While formally radical, this work often disregarded programmatic aspects, political and social concerns, systems of representation, material properties and construction techniques; now much of it seems aesthetically indulgent and too narrowly focused. In the meantime, the architecture, engineering and construction industry was rapidly (and separately) developing software that would streamline the construction process by integrating documentation and project management tools within a single, automatically updated 3-D database: the building information model.

This historic split between design software used for formal experimentation and industry software used primarily for production is giving way to a still emerging area of overlapping concerns. On the one hand, the recent academic trend towards scripted design methodologies – especially genetic algorithms based on a biological model (morphogenesis) – proposes the Darwinian notion of performative fitness as the criterion for selection of form. On the other hand, the profession's further development of specialist software for structural and environmental analyses and simulations, and building information modelling's (BIM's) parameterisation of material properties as well as fabrication and assembly constraints, has elevated practical, technological data to a newly privileged status: it can be produced and considered earlier in the creative process to become an important and integral part of the design concept.

In both of these developments, the usual result has been a systems-oriented, performative architecture that privileges scientific models and practical criteria and favours the development of closed feedback loops of data through embedded tools. While this trend may imbue architectural design with quantitative value and encourage an attractively seamless and efficient design environment, it appears that critical sociocultural aspects of design ('soft' forms of data such as aesthetic strategies, representational aspects, political agendas, sociocultural dimensions and historical material) are often marginalised or left out of these new processes not only because they are qualitative – they are not seen as objective, if quantifiable – but also because they are value laden and potentially politically contentious. At the same time, the primary concern of many practising architects in choosing a BIM package is the potential rigidity and constrained creativity that would be imposed on the design process. As a result many designers still use an intuitive design process and switch to BIM for production, or they collect optimisation data for several factors at the front end of the design process and try to rationalise the often divergent results cerebrally (the optimisation of multiple performative criteria rarely results in a single ideal design outcome and usually requires either the weighting – compromise – or elimination of some factors). Both of these methodologies perpetuate the romantic/rational divide that has plagued architectural design since the Enlightenment as well as the traditional platonic notion of imposing cerebral form on passive matter.[2]

Massive Data
However, there are now indications that a new way of thinking is at hand that could resolve these conceptual divides. In 'The End of Theory: The Data Deluge Makes the Scientific Method Obsolete',[3] Chris Anderson of *Wired* analyses the Google phenomenon and argues that the need to have causal or semantic models is over: the new availability of massive amounts of data (what he calls the 'petabyte'), combined with applied mathematics, supersedes every other tool. Science finds models through a rigorous method of hypothesise/test/confirm in order to go beyond the mere correlation of data and get to the way things really are. But when the scale gets sufficiently large or complex, most scientists concur: the basic laws are not incorrect, they are inadequate.[4] Huge systems (for instance, the dark matter that makes up most of the universe,[5] or even life itself) behave in ways that cannot be deduced in advance by models. Now, in this new Googlien era of massive data and the tools to crunch them, if the statistics say one thing is better than another, the correlation is good enough. 'Who knows why people do what they do? The point is they do it, and we can track and measure it with unprecedented fidelity. With enough data, the numbers speak for themselves.'[6] This approach has already produced impressive and concrete results in physics and biology,[7] and Anderson claims this new type of thinking will shortly become mainstream.[8]

New Tool, New Thinking
Applying this new thinking to architecture and especially BIM opens up a new optimistic world of design possibilities. First, it means freedom from being forced into either the formal indulgence of signature architecture or a hyper-rational mode of performative justifications. And if, as many have said, traditional forms of meaning are bankrupt, then massive data as the new agnostic tool gives us the space to go beyond theories and nostalgic semantics: to have more

1) SLAB IS 'FLIPPED DOWN' ON TO SITE:

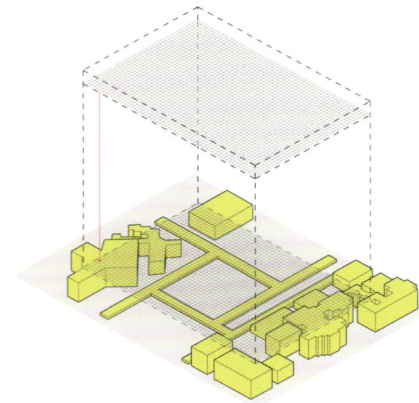

SLAB

EXISTING RUINS

2) THE SLAB GRID TAKES AN 'IMPRESSION' OF THE SITE:

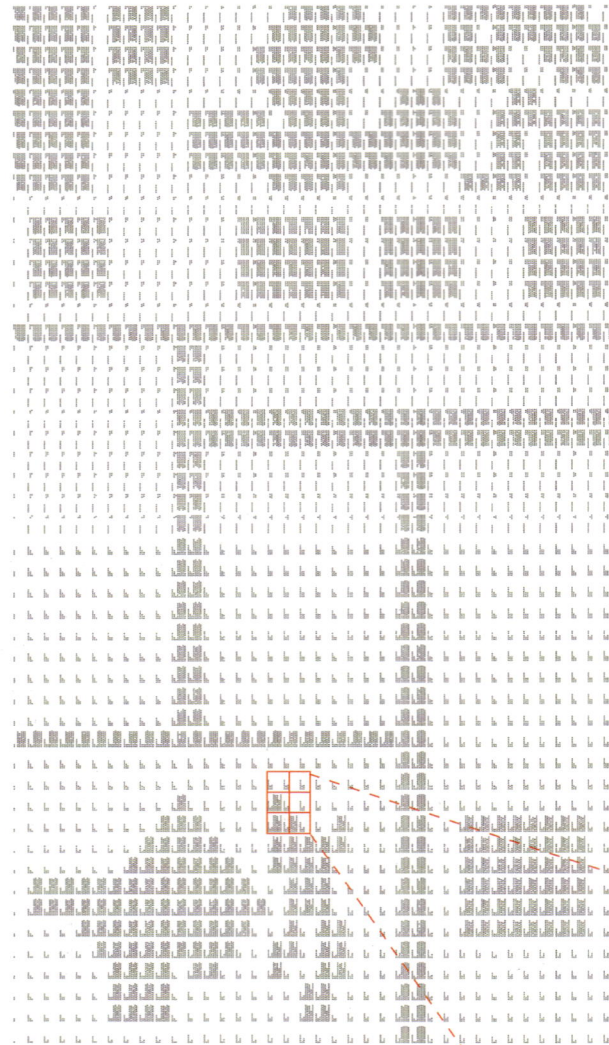

FACADE GRID AS EXCEL SHEET

3) SITE DATA IS REGISTERED ON TO EXCEL SHEET THAT CORRESPONDS TO THE FACADE GRID:

vR (12 , 17)	vR (12 , 18)
96	96
136	144
0	0
0	0
0	0
0	0
0	0
vR (11 , 17)	**vR (11 , 18)**
88	88
136	144
19.99272797	0
30.87765116	0
57.14048338	0
137.0581	0
19.99272797	0
vR (10 , 17)	**vR (10 , 18)**
80	80
136	144
19.99272797	19.99272797
22.88399169	23.89032179
57.14048338	57.14048338
137.0581	137.0581
19.99272797	19.99272797

TYPES OF DATA:

cell position
x position
y position
z position
building height
X pos (central point in XY-plane)
Y pos (central point in XY-plane)
Z pos (central point in XY-plane)

KEY:

SITE

EXISTING BUILDINGS

PROPOSED SLAB BUILDING

OMA Research and Innovation Parametrics Cell, Facade Study for the Slab, 2008

This parametric facade strategy experimented with the soft data of sociocultural material through scripts that articulated the building's internal public zones and registered the site's external urban pattern while incorporating technical functional, structural and material parameters. The series of diagrams shows how the context was translated into data through a process analogous to that of creating a physical impression: 1) the facade grid was conceptually 'flipped down' on to the existing ruins of a site; 2) an impression of each building's location and height was measured; and 3) the data was registered into the corresponding cell of an Excel sheet. A special script was created that keyed this input data to a degree and type of mutation in the facade grid.

options, consider more types of information, and generally be more creative with how we use them. In the Petabyte Age we can 'view data mathematically first and establish a context for it later'.[9] We can incorporate data from many sources, including aesthetics and the sociocultural, political and historical dimensions because massive data is the new meaning. Rather than necessitating the mere reduction of the qualitative into the quantitative, this approach actually creates the conditions for the optimistic potential of emergence: the cumulative effects of complexity and multiplicity may themselves result in the production of new qualities, which in turn allow new relationships between architecture and culture to emerge.[10]

Reconceptualised Role

Such an approach implies that the architect's role should be reconceptualised from that of romantic genius or technological guru to that of multidisciplinary strategist. The new architect is still ultimately responsible for design intent and needs to be able to look at the big picture to decide which factors to parameterise, to give limits to the parameters, assign a weight to each factor and determine the order and method of the information modelling process: in summary to strategise which factors and methods will be used, how they will be applied or generated, and to judge what they contribute. But, just as in the 'new, softer science', algorithms assume the burden of rigorous quantitative methodology and the mind is left 'free to move around the data in the most creative way'.[11] Technically this implies a broader use of creative scripting in the initial stages of design, one that is capable of

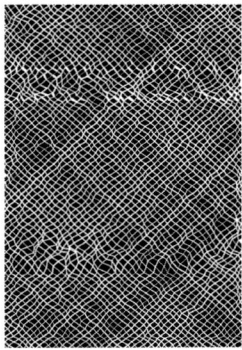

ITERATION_IB_9

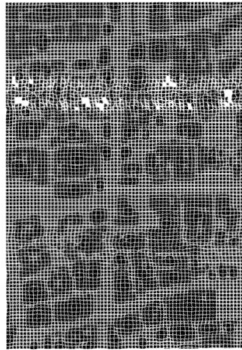

ITERATION_3

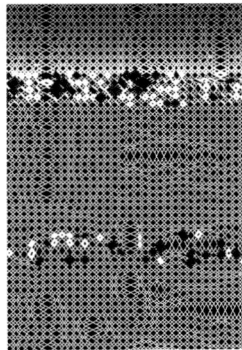

ITERATION_IB_6

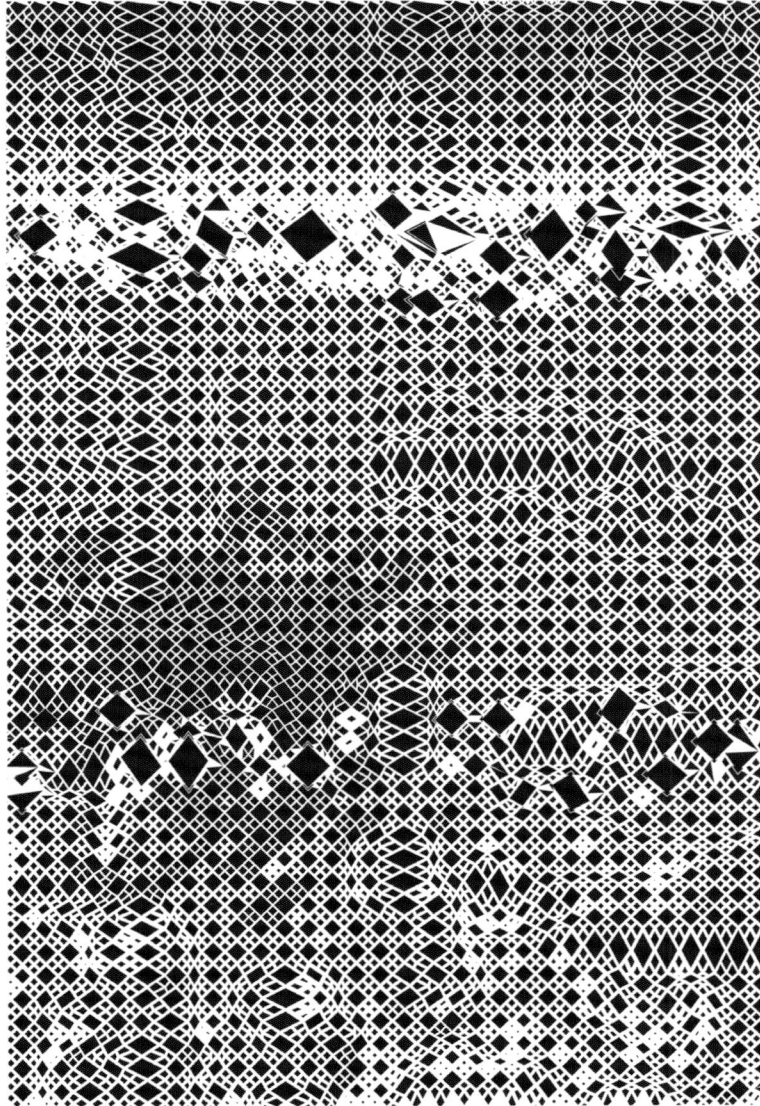

ITERATION_2

MUTATION BASE: CITY MODEL OF OLD DUBAI
LEVEL OF DEFORMATION: VERY HIGH
TYPE OF DEFORMATION: I-DIMENSIONAL
SPECIAL TRANSFORMATIONS: EXTRA MUTATION/DIFFERENT THICKNESS AT THE LOBBIES & ENLARGED
OPENINGS AT THE LOBBIES, RADIAL THICKNESS DEFORMATION AROUND THE LOWER
SKYLOBBY

OMA Research and Innovation Parametrics Cell, Facade Study for the Slab, 2008
A number of data-driven iterations of the facade were generated using different internal and external sociocultural source material, varying degrees (low to high) of mutation, several directional types of mutation (one, two or three dimensions) and a range of functional, structural and material parameters. These iterations, besides articulating the external contextual urban patterns, also indicated the two internal public floors through a different deformation process (a script that randomly 'shook' the grid pattern in these two horizontal zones) and utilised special transformation scripts to increase the level of articulation: extra mutation (different thicknesses) at sky lobby zones, enlarged openings at the lobbies and radial thickness deformation around the lower sky lobby.

recursive growth : based on two orientations and segmented vector lines

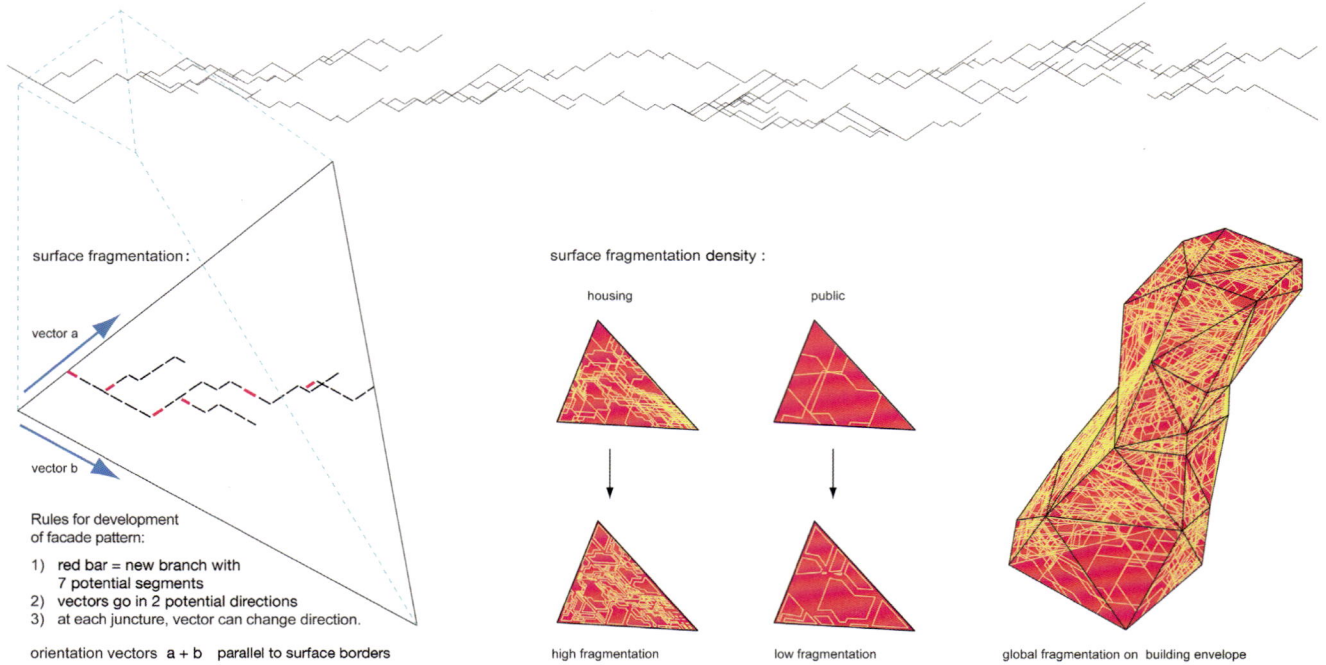

surface fragmentation :

vector a

vector b

Rules for development
of facade pattern:

1) red bar = new branch with
   7 potential segments
2) vectors go in 2 potential directions
3) at each juncture, vector can change direction.

orientation vectors a + b   parallel to surface borders

surface fragmentation density :

housing

public

high fragmentation

low fragmentation

global fragmentation on building envelope

public

hotel

housing

office

office

public

public

north elevation

public

office

public

east elevation

hotel

housing

office

public

west elevation

south elevation

iteration 1

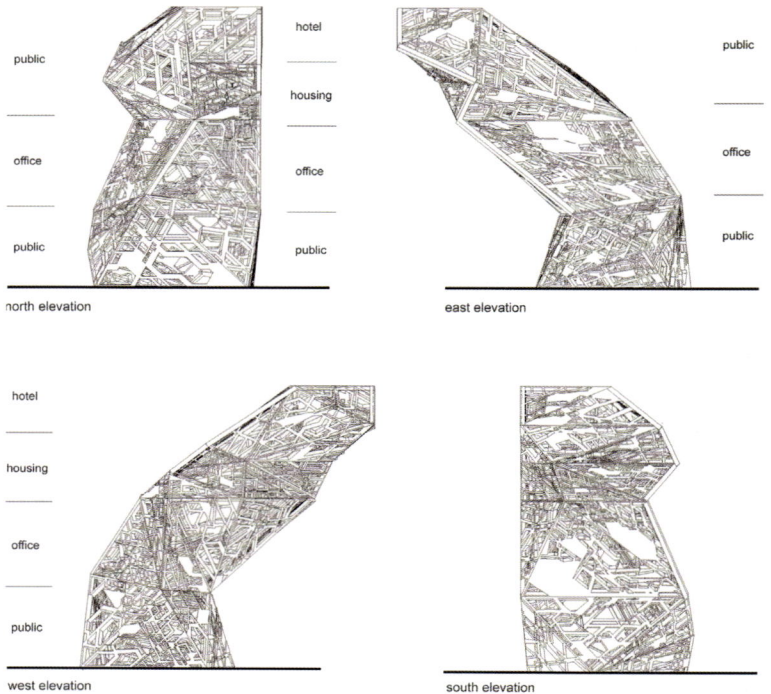

OMA Research and Innovation Parametrics Cell, Study for Phototropic Tower, 2008
The surface geometry of this iteration of the experimental project was generated through a digital
'growth' process based on recursive fractals and their fragmentation. The trajectories of two
possible 'branch' vectors were aligned with the edges of each building facet and the
parametrically limited segmentation of the 'branches' created a varied, but algorithmic,
distribution of structural members. The density and size of the facade members on each facet was
keyed to both the programme behind it and its general position in the building: public
programmes were assigned a more open structure and areas that were cantilevered or carrying
more vertical load required greater structural dimensions for members.

folding in diverse formal strategies and sociocultural data (such as programmatic volumes and adjacencies, the external legibility of programme, site history and so on) beyond the limited scope of, for example, shape grammars. In terms of software, it suggests that rather than trying to develop a single and ideal design environment of embedded tools in a closed feedback loop, we should strive for seamless interoperability – without loss of metadata – between scripting protocols, the more pragmatic stages of BIM and a variety of analytical software.

Conclusion

As proponents of BIM we need to acknowledge the implications of the massive expansion of data and move on from a performative analytical model to a more comprehensive conceptualisation of information modelling that opens up creative options leading to new qualities and relationships, and does not just streamline a process. It should expand the ways we use data rather than merely generating taxonomies or collecting an envelope of constraints. Some of the best designs have been those that have broken the rules and gone beyond technical optimisations or the prescribed constraints of clients and municipalities. Rather than limiting our choices, information modelling can open us up to the new way of thinking and its massive potential. ⚙

Notes
1. George EP Box and Norman R Draper, *Empirical Model-Building and Response Surfaces*, Wiley (New York), 1987, p 424.
2. See Manuel DeLanda, 'Philosophies of design: The case for modeling software', in *Verb Processing: Architecture Boogazine*, Actar (Barcelona), 2002, pp 131–43.
3. Chris Anderson, 'The End of Theory: The Data Deluge Makes the Scientific Method Obsolete', *Wired*, August 2008, pp 106–29.
4. Philip Anderson, 'On My Mind', *Seed*, July/August 2008, p 32. Anderson acknowledges the limits of known laws of physics at larger scales and primitive levels, but also believes computers have limits of error when simulating micro-level interactions.
5. 'Data Set: Dark Matter Observation', *Seed*, July/August 2008, p 32. Only about 5 per cent of the composition of the universe is understood by scientists for certain. The rest is composed of dark energy and dark matter about which little is known.
6. Chris Anderson, op cit, p 108.
7. Ibid, p 109. For instance, J Craig Venter's application of high-speed gene sequencing to entire ecosystems has discovered thousands of previously unknown species of life, based entirely on unique statistical blips rather than on a scientific model or theory.
8. Ibid, p 109.
9. Ibid, p 108.
10. See Gilles Deleuze and Felix Guattari, *A Thousand Plateaus: Capitalism and Schizophrenia*, University of Minnesota Press (Minneapolis), 1987, p 21.
11. Gloria Origgi, 'Miscellanea: A Softer Science,' 3 July 2008. See http://gloriaoriggi.blogspot.com/2008/07/softer-science-reply-to-chris-andersons.html.

DEVELOPMENT OF ITERATIONS:

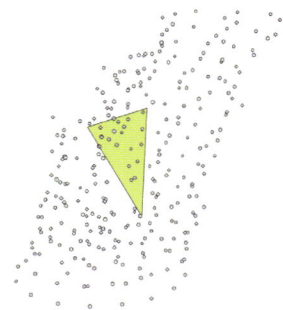

ITERATION_01
SOLAR VS VIEW: 1/1

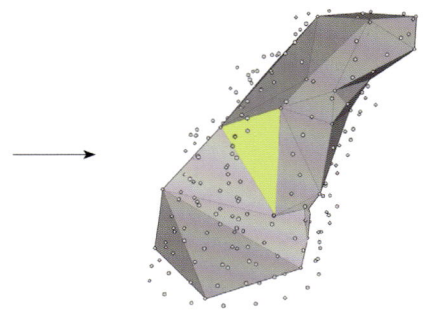

LOW FACET RESOLUTION (20%)

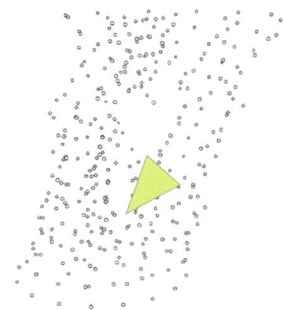

ITERATION_05
SOLAR VS VIEW: 2/1

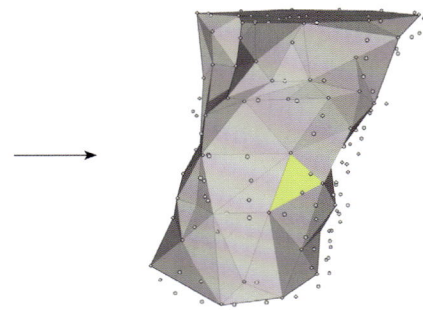

MEDIUM FACET RESOLUTION (35%)

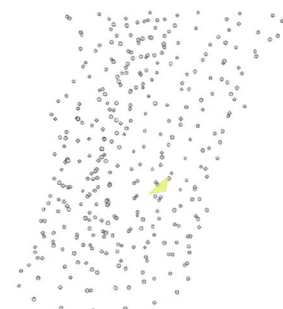

ITERATION_04
SOLAR VS VISTA: 1/1

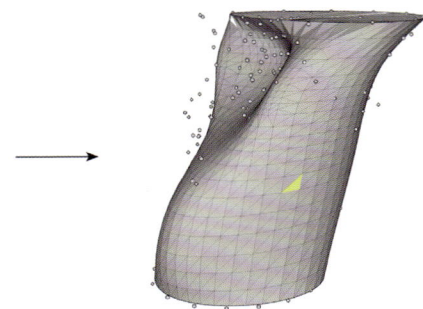

HIGH FACET RESOLUTION (SMOOTH – 100%)

This parametrically designed mixed-use tower project strategically privileges analyses of contextual solar patterns and view foci over normative economic and zoning constraints. Point clouds were developed by processing the following data: sunlight averaged over the year to determine the primary orientation of the building; programmatic data to determine the general floor-plate size and volume; solar insolation data (radiation energy) to determine the first level of deformation of each floor plate; and selected view foci to determine the second level of deformation of each floor plate. Variations of potential building forms were produced by differently weighting the solar and view factors and by the application of varying degrees of faceting (expressed as a percentage) to the point cloud.

New Academic Building for the Cooper Union for the Advancement of Science and Art

New York, 2009

Morphosis

By Martin Doscher

Projective visualisation of the grand stair, central atrium mesh and skip-stop stairs. The information model used for this rendering is the same one used for technical drawings. A whole range of non-geometric information feeds into, and is generated from, the design process. Morphosis' aim has been to allow front-end control over a range of project inputs, such as occupation density, cost and pace of development, to dynamically drive a series of possible development scenarios.

The new Cooper Union building under construction, showing the multiple layers such as the outermost sunscreen layer (left), glazing (centre) and the exposed structure (top). Contextual modelling and analysis is one area that presents opportunities for developing new approaches through information modelling. At Morphosis, issues of urban context are particularly important in the idea-generating processes of a project. Information modelling allows contextual cues to move from implicit inputs to explicit inputs. With this explicit knowledge the firm's designs can be directly responsive, and responses can be measured in a much shorter feedback loop.

Detail highlighting the resolution of the handmade glass-fibre reinforced gypsum joints. Through this collaborative process, which takes advantage of the plasterer's expertise, both the architectural intent and a high degree of precision can be achieved without the need for costly CNC-based processes.

This new academic facility is a stacked vertical piazza contained within a semi-transparent envelope that articulates the classroom, laboratory and art studio spaces. It is organised around a central atrium that rises the full height of the building. This connective volume, spanned by sky bridges, opens up view corridors across Third Avenue to the foundation building.

The configuration of the interior spaces encourages interconnection between the school's engineering, art and architecture departments. To make these connections, the circulation system relies upon several staircases that pass through the central atrium and are augmented by a skip-stop elevator system. The elevator system connects only the ground, fourth and seventh storeys, with stairs from the elevator lobbies on these floors serving adjoining levels. A grand stair connects the ground floor to the first, second and third storeys. Focal staircases individually connect from the fourth floor down to the third and up to the fifth, and likewise connect the seventh floor down to the sixth and up to the eighth.

Department amenities – including meeting rooms, social spaces, seminar rooms, wireless hubs and computer drop-in centres -- are located in the fourth- and seventh-storey sky lobbies that surround the atrium. Classrooms, offices and other functions are distributed in the adjoining floors above and below. The prominence of the grand stair, both visually and in terms of functionality, encourages students, faculty and visitors to use and congregate in this space; in practice, half will use the stairs as their sole means of circulation. These key social spaces thus become the places where informal education takes place.

The physical and visual permeability of the building helps integrate the college into the neighbourhood. At street level, the transparent facade invites passers-by to observe and take part in the intensity of the activity within. Many of the public functions (including retail space and an exhibition gallery) are located at ground level, and a second gallery and a 200-seat auditorium are also easily accessible from the street.

The open, accessible building is a great example of sustainable, energy-efficient architecture. A steel-and-glass skin improves its performance through the control of daylight, energy use and selective natural ventilation. This double-skin system allows for heightened performance and dynamic composition on several levels: the operable panels create a continually moving pattern, provide surface variety on the facade, reduce the influx of heat radiation during the summer, and give users control over their interior environment and views to the outside.

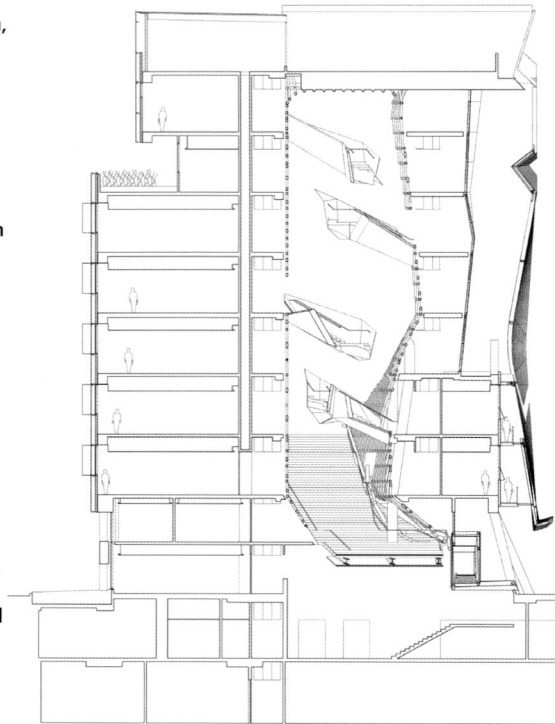

Transverse building section. The atrium carves out a social space through the building, working in conjunction with the skip-stop elevator system, which serves the ground, fourth and seventh floors. Circulation to other floors passes through the atrium. The geometry of the atrium is carved out on to the facade beyond the glass-fibre reinforced gypsum mesh, revealing a view of the foundation building and the city beyond.

Third-floor plan. Laboratories and offices are organised around the perimeter and are linked by the more expressive atrium that forms a connective social space through the centre of the building. The New York City Building Code generally does not permit nine-storey atria, thus the design proposal required performance simulation for smoke evacuation studies as well as analytical studies to satisfy the code. The plan shows the grand stair to the right, and a staircase in the centre of the atrium leading up to the skip-stop elevator lobby on the fourth floor. The atrium opens up like an aperture to a large glazing panel on the right-hand side of the plan.

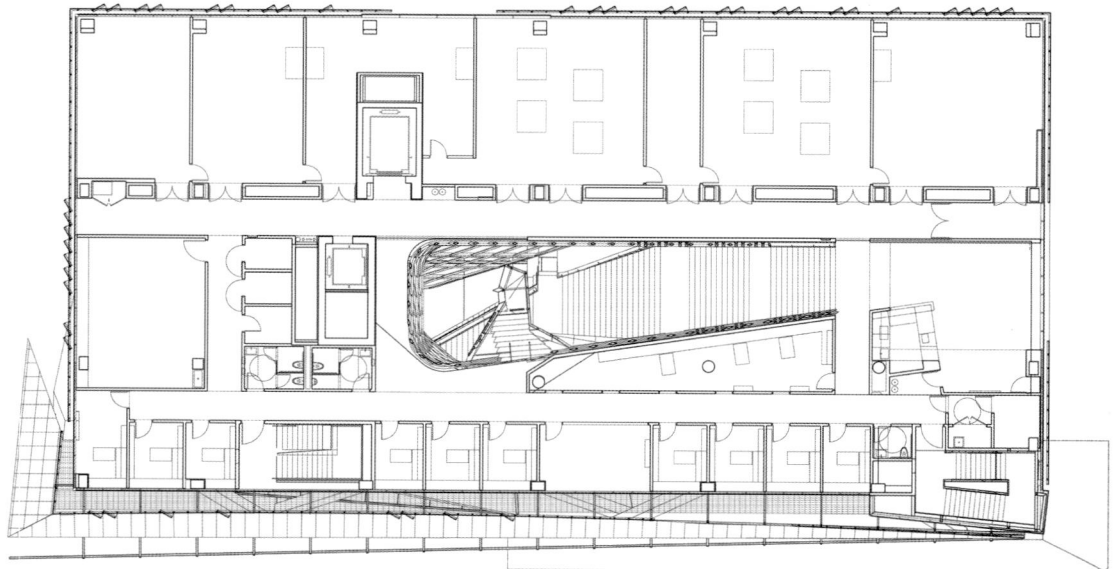

9th floor

8th floor

r = 5'

8" 10 1/4" 3'-7 1/2 10 5/8"

3'-6"

5'-9 5/8" 5'11 3/8

r = 5'
3'-8 3/4

2'-8 7/8

1'-4 1/8" 2'-8 1/2"

DEVELOPMENT OF ATRIUM VOLUME

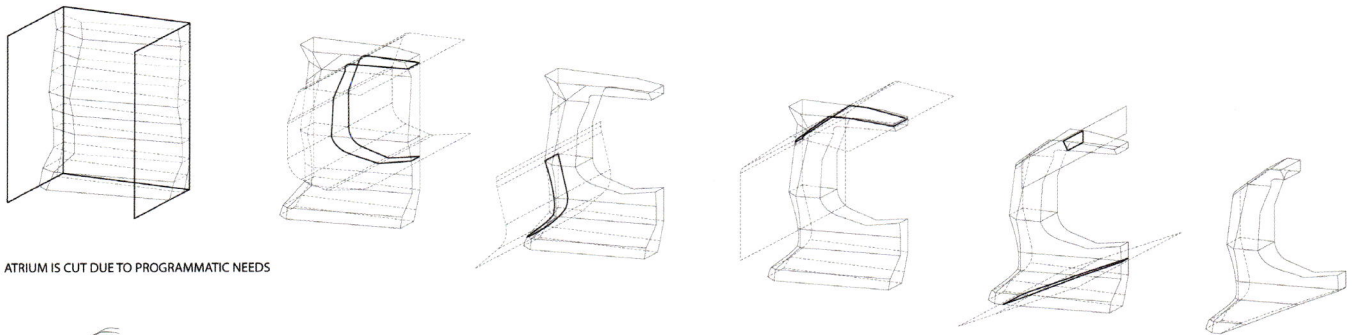

Atrium mesh construction mock-up diagram. Design development drawings are generated so that a discussion with fabricators can occur as projects are moved into detailed design. At Morphosis, the design is married with the production method to control costs as well as to ensure the project intent is maintained. Mock-up diagrams which extract a local portion of detail to be studied are critical to this process. This also allows for collaborative relationships during the design and fabrication stages.

ATRIUM IS CUT DUE TO PROGRAMMATIC NEEDS

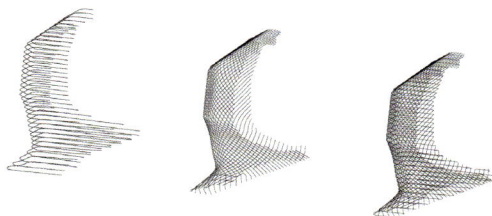

CREATION OF ATRIUM MESH

Rationalisation of the atrium surface geometry from a vertical void within the building's mass. Programmatic and circulation needs dictated the atrium form and allowed for the refinement and formal development of the atrium mesh. Δ⅁

Text © 2009 John Wiley & Sons Ltd. Images © Morphosis

The $300,000/Year Architect

Could building information modelling (BIM) shake up not only design and delivery processes, but also be set to redefine the construction hierarchy? Could it shift the architect's position from near the bottom of the current food chain to one nearer the top? **Urs Gauchat,** Professor and Dean of the School of Architecture at the New Jersey Institute of Technology, outlines the architect's place in a complex future where there is an increasing demand for buildings but a demise in available materials and conventional energy sources.

The idea of the architect as choreographer who makes traditional approaches to dance needs to be expanded to allow for form-giving based on parametric design and non-traditional ideas. At this point architecture no longer needs to be restricted to an extrapolation or interpolation of historical precedents. With information modelling, architects can simultaneously be much more connected to design while understanding construction, and bringing together design and making like never before.

The impact of building information modelling (BIM) and integrated project delivery (IPD) is likely to create more opportunities and more pitfalls than the CAD revolution has managed to over the last 25 years. It is not yet evident whether BIM will lift architects to new heights or whether it will lead to further marginalisation. Unlike CAD, its transformative power on the profession will be far swifter and thus more consequential.

This positive trend will provide the context for BIM for at least the next 25 years. The increase in building volume will be driven by the rapidly expanding economies of China and India, and by an unprecedented rate of construction elsewhere. Even given the financial outlook in the US and around the globe at the time of writing, the rate of construction will, over the coming years, be unprecedented. Currently, all of the buildings in the US

account for approximately 230 billion square feet (21.4 billion square metres). It is anticipated that by 2030 there will be a need for an additional 106 billion square feet (9.8 billion square metres) of new construction as well as 97 billion square feet (9 billion square metres) of reconstruction.[1] This is the equivalent of 90 per cent of all buildings in the US today.

Similar unprecedented rates of growth are predicted throughout the world. The worldwide need for building will create a scarcity of resources that will bring about radically different building design and delivery methods. The materials, methods and techniques employed now will have to evolve in response to finite commodities such as iron, ore, coal and bauxite. In addition, staggering amounts of energy will be consumed in the production of building materials such as cement, glass, steel and aluminium. Other building materials, including plastics, are petroleum based.

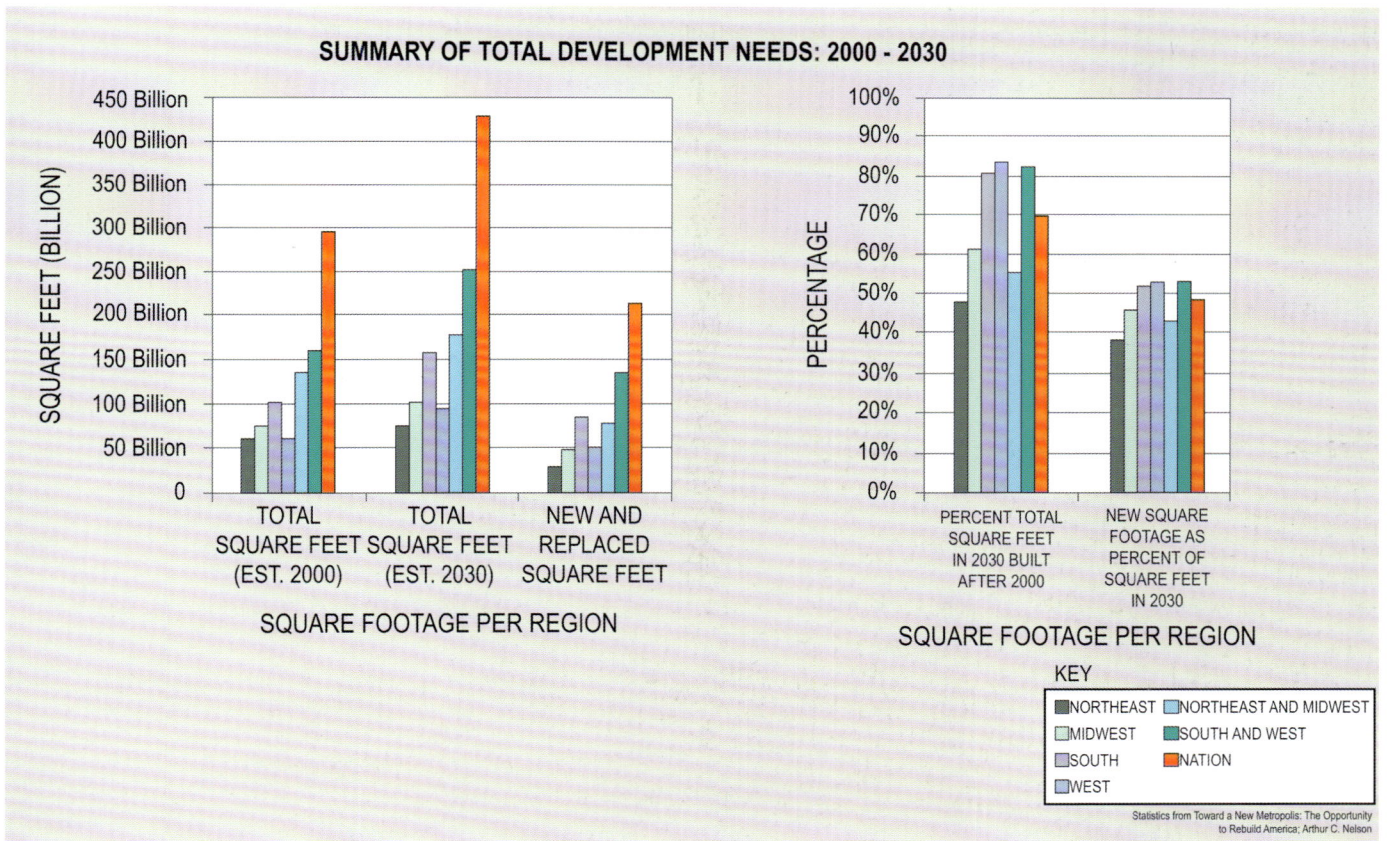

SUMMARY OF TOTAL DEVELOPMENT NEEDS: 2000 - 2030

Despite the present economic woes, the long-term prognosis for construction remains positive. The long-term trajectory is, of course, subject to cyclical moves in the economy. Nevertheless, the long-term prospects are driven by strong underlying fundamentals and, despite economic fluctuations, will continue to trend upwards. Data from Arthur C Nelson, *Toward A New Metropolis: The Opportunity to Rebuild America*, The Brookings Institution (Washington DC), 2004. Graphic based on Appendix Tables, pp 26–41.

Buildings are already responsible for approximately half of the world's CO_2 emissions, as well as disproportionate amounts of energy, water, refuse and pollution. All these factors will add to the pressure to develop new approaches and innovations that conserve not only energy, but also finite materials in the manufacture, construction, operation and recycling of buildings. This new context will sorely test the ability of the architectural profession to adapt and adjust to a new business climate, and BIM will be one of the keys to this necessary metamorphosis of design and construction.

BIM not only provides the possibility of seamless communication between all the participants in the building process, but also holds the promise of real-time links to various live databases. This, in turn, will have a significant impact on the decision-making process as design projects are developed, with informed decision-making processes leading to more highly informed design proposals.

The present design and building process can be seen as a Darwinian food chain. At the top of the food chain are investors, private equity firms, hedge funds and financial institutions. Somewhat lower down, yet still near the top, are developers and owners. Still lower are contractors, owners' representatives and construction managers, followed by architects, engineers and allied professionals. Near the bottom of the food chain are construction workers and others engaged in manual labour. By enhancing the architect's ability to create value, BIM could radically change the profession and wrench it from its present trajectory of marginalisation, eventually opening the way for architects to move further up the food chain.

Parallel Developments
A parallel, but equally relevant, development is the quiet revolution that has taken place in the real-estate business. This is due to a radical shift in the nature of building clients and sponsors. The days when aristocratic and benevolent patrons were the major sponsors of architecture are now in the distant past. Also waning are individual entrepreneurs with edifice complexes. These traditional clients are increasingly being displaced by private equity firms, hedge funds, real-estate investment trusts (REITs) and other financial institutions. To these new entities, real estate is just another financial vehicle: they are driven by sophisticated analyses rather than ego or pride of ownership; they are shrewd and constantly compare real-estate investments with alternative and potentially more productive uses of capital. Their investment strategies centre around three major variables: yield, safety and liquidity. Recently, returns on investments in commodities have

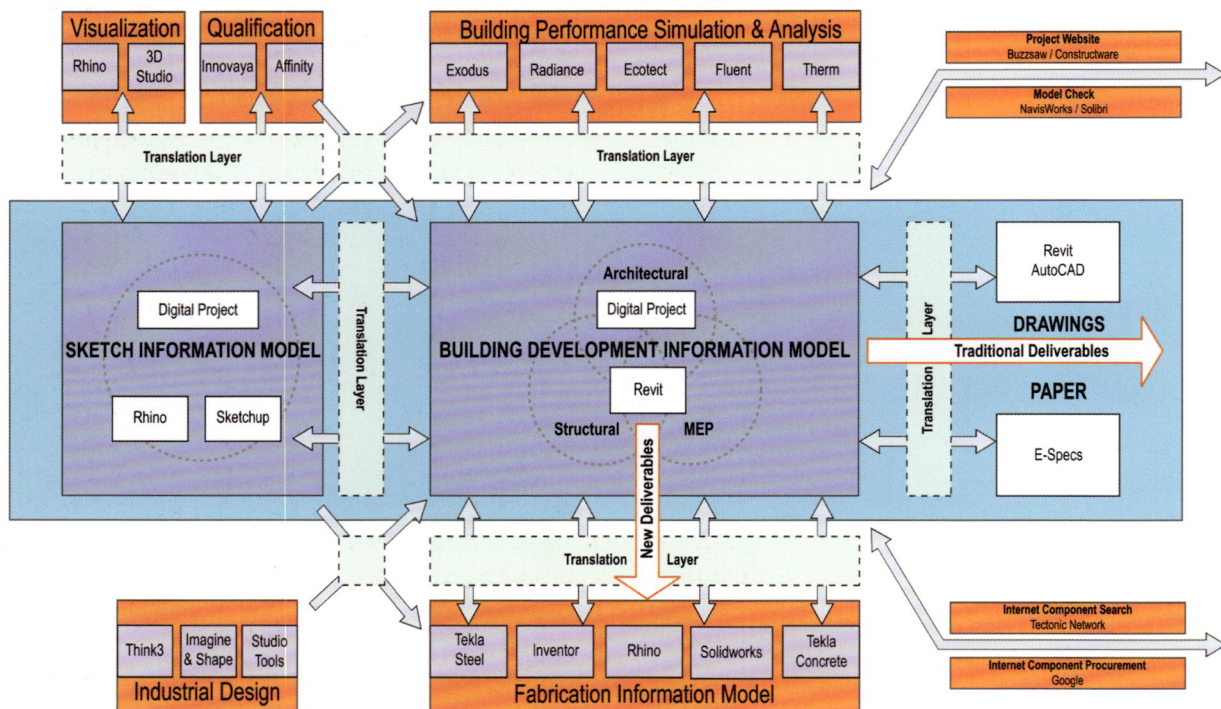

Image Courtesy of Skidmore Owings + Merrill

Skidmore, Owings & Merrill (SOM), Building Development Information Model, New York, 2008
Building information modelling (BIM) platforms such as Revit and Digital Project allow a 3-D virtual computer model to be linked to any variety of databases used for the scheduling of components or the costing of those components. BIM can dramatically expand decision making and information flow, an approach that produces a much higher degree of accountability and predictability, and also raises the question of who is in charge of the process, who owns the virtual model, who is taking the risk and who reaps the rewards. The chart here, developed by SOM, illustrates this point.

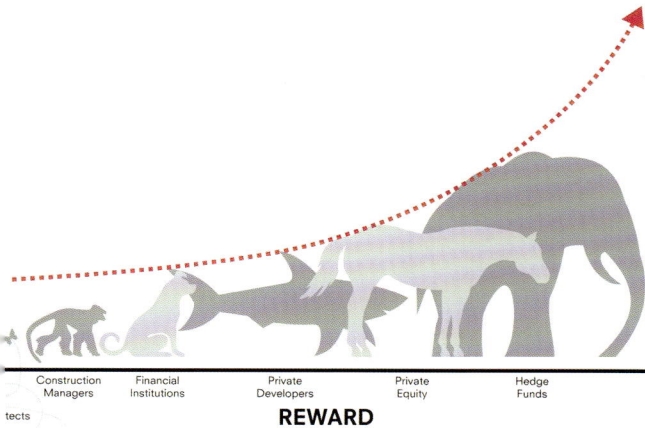

SHoP Architects, Architectural Practice
Risk and Reward, New York, 2007
The Darwinian food chain in terms of
reward compared with the intellectual
contribution to the design and building
process. BIM could finally open the way for
architects to move further up the food chain.
Those lower on the food chain, including
architects, have traditionally been seen as
non–essential to the construction process
and therefore easily replaced.

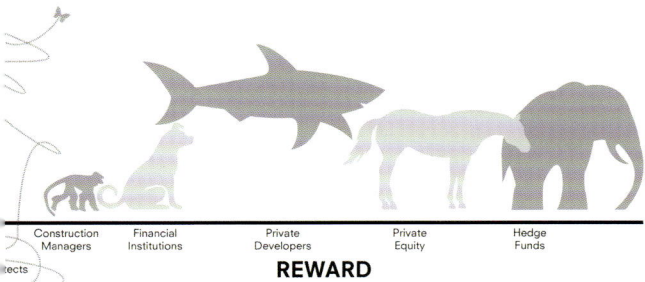

By enhancing the architect's ability to create value by
linking decisions made during the design phases to
construction and assembly operations during the
construction phases, BIM has the potential to radically
change the profession and to wrench it from its
present trajectory of marginalisation. Architects may
then be able to re-establish themselves as central to
the construction process.

been promising, and consequently large amounts of capital has shifted
to commodities from less liquid and lower-yield investments such as
real estate. Capital reallocations such as this are triggered by mega-
trends as well as local economic fluctuations. Since capital does not
have any allegiance to sector nor location, it will inevitably seek to find
the highest return with the least amount of risk.

Given this situation, what is the role of information models and IPD?
BIM makes it possible to see the effects of decisions in terms of cost,
energy efficiency, material availability and other project expenditures.
It is therefore of particular interest to this new kind of client. The fact
that a building can be run and tested via an electronic model is akin to
Boeing creating a virtual 777 and testing all the systems electronically
prior to it going into full-scale production.[2] BIM removes uncertainty
and therefore reduces risk. Owners, financial institutions and investors
alike have a pathological dislike for unknowns and concomitant risks,
thus the added value provided by architects using BIM justifies higher
fees and even their possible participation in the economics of building
development projects.

Information Models in Design Practice

It is important to keep in mind that BIM is merely a tool, albeit a
transformative one. Therefore, contrary to a common misconception,
BIM should have no negative impact on the quality of design. Indeed,
the ability to do cost and benefit analyses during the design process
should encourage more daring and unconventional design as alternative
and far-fetched solutions can be evaluated on an equitable basis.

Many view BIM primarily as a software program. This is based on
the assumption that 'CAD jockeys' would merely be replaced by 'BIM
jockeys'. Instead, the use of BIM to its full potential will require very
high levels of skill, experience and judgement. The traditional
draughtsman or junior architect of yore is not in a position to
simultaneously design and evaluate a myriad interlocking decisions.
The integration of design decisions, cost implications and
construction feasibility requires an inordinate ability and a highly
elevated level of responsibility. The activities of such an architect
can be compared to piloting a modern commercial jet instead of a
single-engine private plane. This new professional will have a much
greater impact on the quality of design, constructability and cost.
Individuals capable of performing such a complex and demanding
role are rare and will therefore be sought after. The added economic
value they can provide should elevate their salaries to the equivalents
of those of top lawyers and doctors.

The impending changes have some interesting effects on the nature
of practice. Architectural practice is in the process of bifurcating into
the adherents to traditional models and those who take advantage of
technological change. The technologically more advanced firms, such
as those represented here, are likely to grow further and further apart
from the pack; they will be capable of providing much higher and more
comprehensive levels of service and raising the expectations of clients.
It is likely that it will become increasingly difficult for traditional firms
to compete. In other words, the penalty for resisting change will
increase dramatically over time.

BIM has for some time played a significant role in non-building applications, and other markets have been ahead of the building design and construction industry in taking advantage of information modelling. It has allowed engineers and scientists to simulate conditions to virtually fly planes and run refineries prior to construction. Going beyond depiction, information models, such as this fuselage of a Boeing 777, are operative and bring a new level of sophistication and informed decision making to design and building.

Information Models in Architectural Education

Just as the profession is likely to bifurcate, so too will the architectural schools: into the more traditional establishments that will continue to foist an educational programme on students that was modelled on the teachers' own educational experiences, and the more progressive schools that are able to make the necessary investments and faculty adjustments to educate a new breed of architects. These two trajectories are also likely to grow further and further apart. For each school the key question is how to position itself vis-à-vis other schools and vis-à-vis a changing profession.

In principle, much of traditional education in architecture was based on an emulative model that has changed little since the Beaux Arts period. It was predicated on the idea of an apprenticeship to a master. A student's education was based on building blocks of knowledge from authoritative sources taught by authoritative figures. The challenge of architectural education today is how to produce a new kind of professional with distinctly different skills sets, attitudes and knowledge.

The current conundrum in architectural education pits the traditional pedagogical framework, which relies heavily on personal visual judgements, theories of composition and proportion, against parametric design, which combines performative criteria with algorithm-driven design. Parametric design offers solutions that can neither be imagined nor judged by today's subjective-aesthetic standards. BIM has a place in both of these approaches, though it is more integral to parametric design.

Another aspect of the educational equation is the changing nature of architecture students. Millennials (that is, students born between 1980 and 2000) have different qualifications and interests compared to previous generations of students. But more importantly their brains seem to be wired differently – the result of exposure of their brains to digital information during the early developmental stages and during periods of particularly active synapse formation.[3]

A changing professional landscape will require changing emphases in architectural education. New imperatives will bring different professional attributes to the fore and displace many time-honoured skills, attitudes and knowledge. Despite the most far-sighted and informed goals, each institution is operating under many real or imagined constraints. The education of the complete professional of tomorrow must emphasise some attributes that depart markedly from the past, for example:

- Judgement is probably the most important yet elusive characteristic of a complete professional. With almost infinite amounts of information, judgement is needed to discern which information is relevant and which is extraneous.
- Teamwork will be required to design and implement complex building projects. Increasing time pressures and economic realities make it imperative that teams have access to the necessary expertise to make informed decisions. The most challenging aspect of teamwork, however, is to foster rather than inhibit creative design and design excellence. Teamwork implies neither consensus nor collective decision making. Instead, design teams require inspired leadership to avoid being mired in mediocrity.

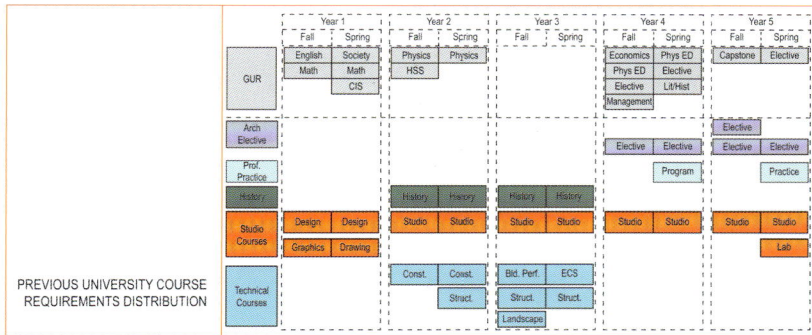

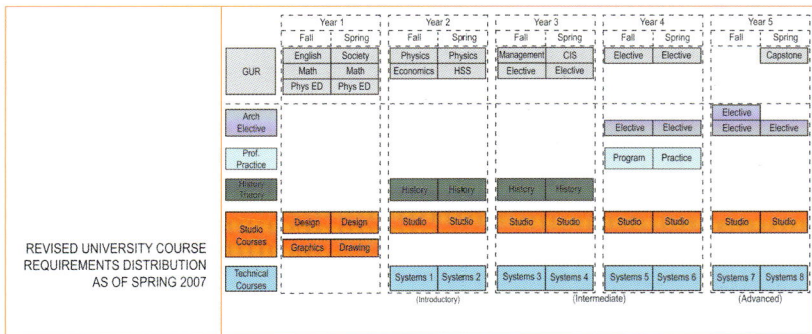

Curriculum of the School of Architecture at the New Jersey Institute of Technology. Instead of a traditional framework of dispersed 'technical' courses such as structural fundamentals or environmental and enclosure systems, the school recently implemented a series of eight building systems courses that seek to expose relationships in the design and coordination of building systems (plumbing, structural, mechanical and so on) within an architectural context. Coupled with specific design problems consistent with the study year, the content imparted can be more successfully linked through information modelling to become the common platform for studio and systems courses. It is designed to reflect the needs of the profession 15 years hence, and is intended to foster the key attributes of the complete professional of tomorrow.

PREVIOUS UNIVERSITY COURSE REQUIREMENTS DISTRIBUTION

	Year 1 Fall	Year 1 Spring	Year 2 Fall	Year 2 Spring	Year 3 Fall	Year 3 Spring	Year 4 Fall	Year 4 Spring	Year 5 Fall	Year 5 Spring
GUR	English / Math	Society / Math / CIS	Physics / HSS	Physics			Economics / Phys ED / Elective / Management	Phys ED / Elective / Lit/Hist	Capstone	Elective
Arch Elective									Elective	Elective
Prof. Practice							Elective	Elective	Elective	Practice
History			History	History	History	History				
								Program		
Studio Courses	Design / Graphics	Design / Drawing	Studio	Studio	Studio	Studio	Studio	Studio	Studio	Studio / Lab
Technical Courses			Const.	Const. / Struct.	Bld. Perf. / Struct.	ECS / Struct.				
					Landscape					

REVISED UNIVERSITY COURSE REQUIREMENTS DISTRIBUTION AS OF SPRING 2007

	Year 1 Fall	Year 1 Spring	Year 2 Fall	Year 2 Spring	Year 3 Fall	Year 3 Spring	Year 4 Fall	Year 4 Spring	Year 5 Fall	Year 5 Spring
GUR	English / Math / Phys ED	Society / Math / Phys ED	Physics / Economics	Physics / HSS	Management / Elective	CIS / Elective	Elective	Elective		Capstone
Arch Elective									Elective	Elective
Prof. Practice							Elective / Program	Elective / Practice	Elective	Elective
History			History	History	History	History				
Studio Courses	Design / Graphics	Design / Drawing	Studio	Studio	Studio	Studio	Studio	Studio	Studio	Studio
Technical Courses			Systems 1 (Introductory)	Systems 2	Systems 3 (Intermediate)	Systems 4	Systems 5	Systems 6	Systems 7 (Advanced)	Systems 8

- Technical skills are also essential components of a complete professional. BIM and IPD are among the critical techniques that will enable the next stage in the evolution of the profession. The need to communicate effectively using different BIM platforms will place increasing emphasis on interoperability and agreement on common standards.

- Intimate knowledge of the design and building delivery process is needed in order to be the most effective participant. Particularly in the US, the current constellation of professional roles and contractual relationships are a preamble to conflict and acrimony. A realignment of interests in which all participants share in the success or failure of a project provides a more effective and profitable *modus operandi*. As the boundaries between professional roles become further eroded, a full realisation of the potential of IPD will result in more productive and profitable collaborations.

- General education must not be marginalised. A rudimentary understanding of business principles, psychology, social science and traditional liberal arts provides an effective remedy against the myopia of many professionals. Architectural education must not give in to pressures to trade off general education in favour of professional subject matter. Much professional knowledge ages quickly, whereas good general education will always provide a sound framework for lifelong learning. Moreover, a general education provides an excellent opportunity to hone skills in communication and reasoning.

- Talent comes in many forms. In complex buildings there is a continual give and take between team members, who use their respective talents. Although the designer is still first among equals, he or she may no longer necessarily lead the design process. Complementary expertise and a range of talents are the prerequisites to a successful design solution.

The way in which these attributes are addressed in the education of an architect will vary from school to school. In each case, a close collaboration between the profession and the school will ensure that the respective educational model is properly calibrated. Changes in the profession must be the result of collaboration between educational institutions and architectural firms; they are dependent on each other and can thrive only through constant give and take.

It is crucial that educational models be developed that unleash the creative impact of inspired design using the full arsenal of new techniques. Architecture is at an exciting crossroads. Both the profession and education need to work closely together in the pending metamorphosis to ensure that architecture will have the best days ahead of it, rather than behind it. What an exhilarating prospect. ∆

Notes
1. Arthur C Nelson, 'Toward A New Metropolis: The Opportunity to Rebuild America', Brookings Institution Report, 2004.
2. Guy Norris and Mark Wagner, *Boeing Jetliners*, Zenith Imprint – MBI Publishing (Minneapolis), 1996, p 21.
3. Maryanne Wolf, *Proust and the Squid: The Story and Science of the Reading Brain*, HarperCollins (New York), 2007.

Al Hamra Firdous Tower

Kuwait City, Kuwait, 2008-

Skidmore, Owings & Merrill

By Gary Haney

At night, the central void of the tower is lit from within. As the envelope turns, the light fades away, emphasising the three-dimensional quality of the tower and its internal space.

The lobby structure is one of the most prominent elements of the project. Designed as a cross-bracing structural system, this 20-metre (65.6-foot) high formal entrance to the building acts as a light filter during the day and reveals its silhouette at night.

Located in the centre of Kuwait City, the Al Hamra Firdous Tower is a 412-metre (1,352-foot) high skyscraper housing a 20-metre (65.6-foot) tall lobby, two sky lobbies, a sky garden, six service floors and 62 office floors. Currently under construction and due for completion in 2010, the award-winning structure is designed to maximise views of the gulf and minimise solar heat loads. Its unique sculptural form was generated through an algorithmic design process. This was subtractive, similar to chiselling out the parts that are not needed from a block of stone. The 60 x 60 metre (197 x 197 foot) building footprint was first extruded to 412 metres (1,352 feet). The centre of the square was then hollowed out to define a 13.5-metre (44-foot) lease span, creating a doughnut-like form. To achieve the target 2,200 square metres (23,680 square feet) floor area, one quarter of the doughnut was removed, starting from the southwest segment at ground level and gradually rotating counterclockwise to the southeast segment at the top to maximise views and minimise solar heat loads. To accentuate the spiralling of the void, the southeast tip of the building was raised to become the tallest point.

The perception of movement around the tower is the result of the spiralling trajectory of the removed floor plates. Designing the subtraction process of the object, rather than the object itself, creates a phenomenal artefact that has a perpetual dialogue with its missing counterpart.

The simple design logic of the tower, which can be summarised as one continuous line, creates complex, double-curved, sheer concrete flare walls at the cut edges of the doughnut. These are sheeted with rainscreen walls, the stone panelling for which, following the reference geometry, was rationalised through a computational process using point clouds and scripting. This algorithmic approach to design created a rule-based framework for a series of interconnected and interdependent processes to occur, giving a rational, almost scientific, rigour to the design process.

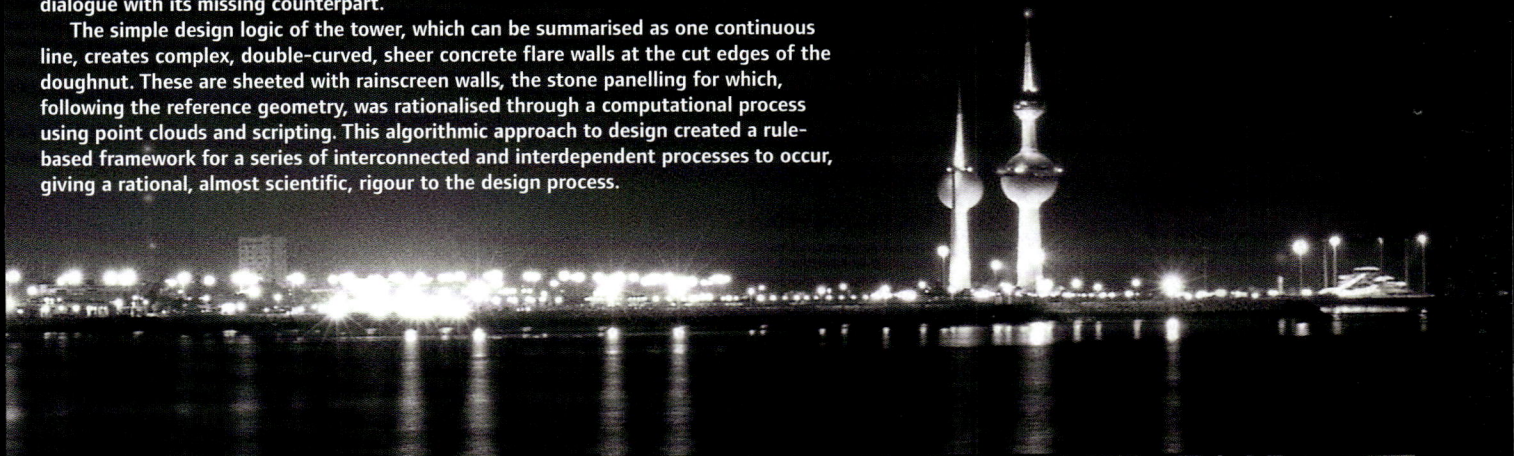

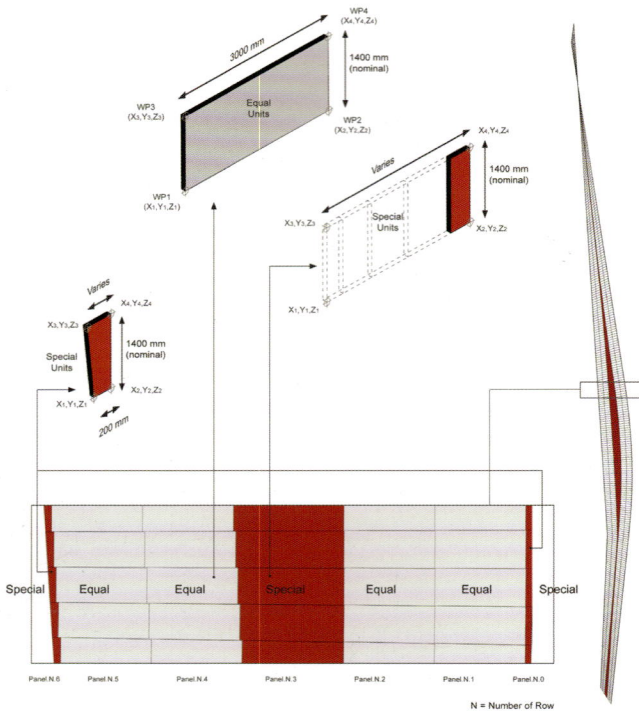

Flare–wall subdivision. Despite the complex geometry, the optimisation of the panelling process yields a highly economic result of 94 per cent identical stone panels.

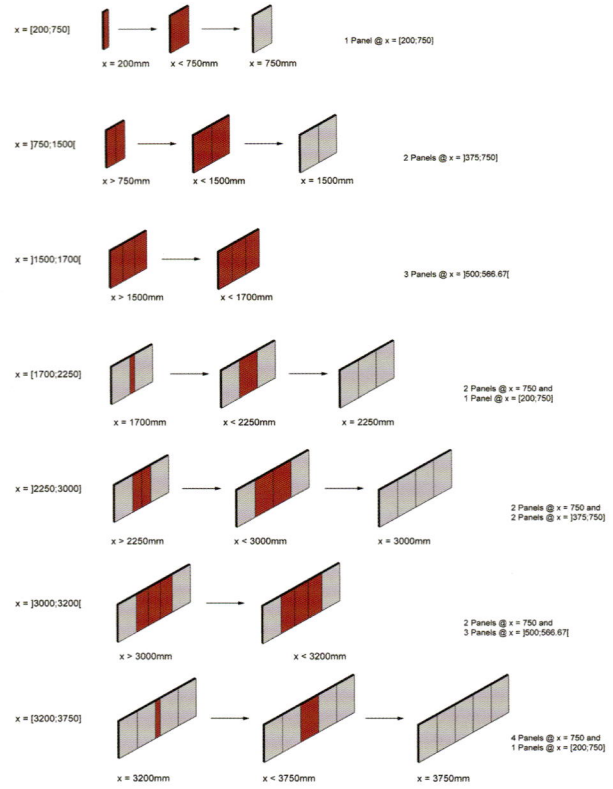

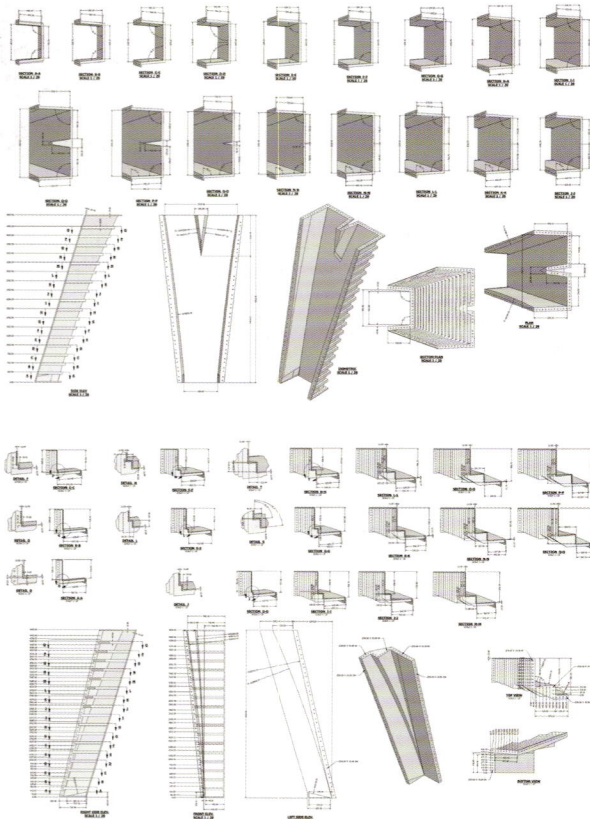

Lamella model. The spatial and aesthetic qualities of the lobby and its structural members were tested and confirmed using 3-D printing technology.

Formwork shop drawings. The shop drawings were derived from the 3-D model of the lobby and show the multiple plan cuts along the length of the structural members.

Construction photograph and digital model. The 3-D lamella model was instrumental in the design documentation process. The accurate three-dimensional description of the structure allowed for its realisation within the constraints of a challenging time frame.

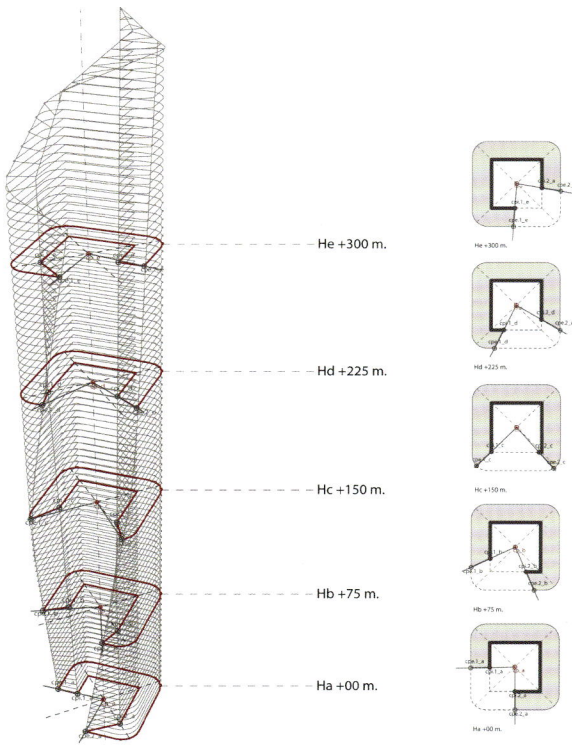

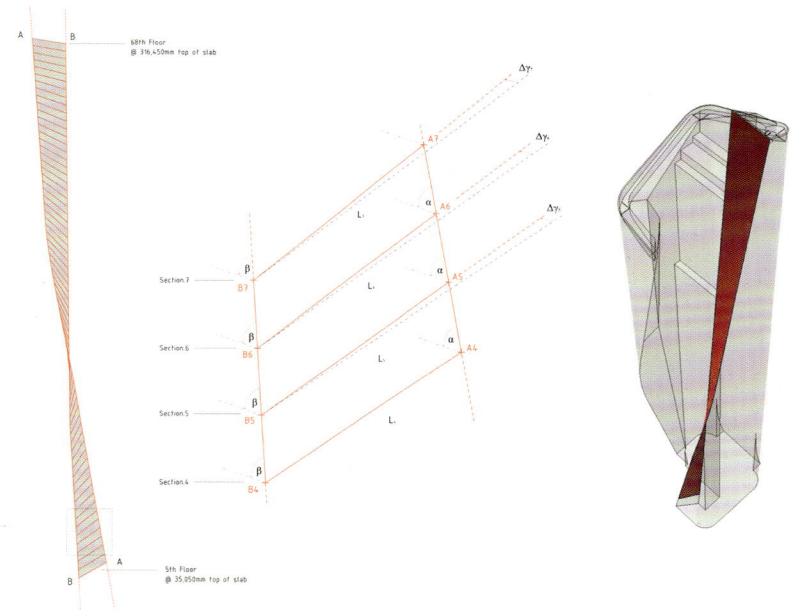

He +300 m.

Hd +225 m.

Hc +150 m.

Hb +75 m.

Ha +00 m.

Geometry formation diagram. The form of the tower was generated through an algorithmic design process. This process is subtractive, similar to chiselling out the parts that are not needed from a block of stone.

The shape of the hyperbolic flare walls is the result of the subtractive operation in the form–generation process. Geometrically, these walls are defined as hyperbolic paraboloids. △

Unified Frontiers
Reaching Out with BIM

Coren Sharples is a partner of SHoP Architects in New York where building information modelling (BIM) has become integral to the fluid promotion of communication between contractors, consultants and clients on a project. This is exemplified by two projects in the city: the reconstruction of Rector Bridge, where BIM enabled SHoP to work seamlessly with a firm of custom boatbuilders and other contractors, and the speculative development at 290 Mulberry Street, where it became a significant means of disseminating key data to the client for marketing and financial planning.

Architecture has always been packaged in eras of style. Today's culture of technology allows us to break from this model and look towards a future of performance-based design. This is not to say that aesthetics, or stylistic trends, disappear or become unimportant, but that the designer gains creative freedom when the end result is measured against performance criteria rather than rules of style.

The education of the architect is one of problem solving. In order to be open to new and unexpected ways of thinking, it also helps to have a broad education, with an awareness of the world outside the profession. This was the model during the Renaissance, which produced the architect as master builder. The partners and many of the staff at Sharples Holden Pasquarelli (SHoP) have undergraduate degrees in diverse subjects such as history, fine arts, English, law, engineering, economics, finance and biology. This impacts the way the firm thinks, solves problems and engages with others including engineers, clients and contractors. The practice is inspired by the example of Burt Rutan who, with a group of young engineers, was able to formulate a strategy for the design of SpaceShipOne, which won the $10 million Ansari X-

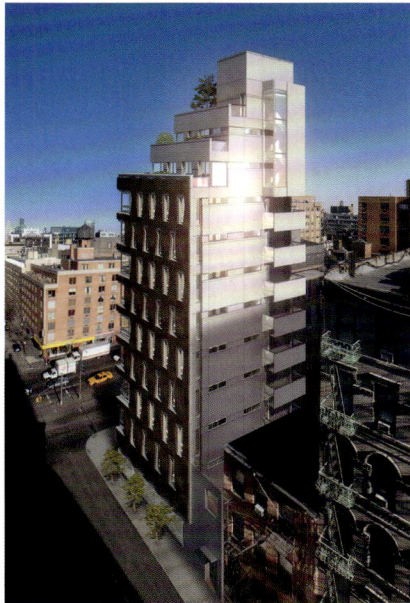

SHoP Architects, 290 Mulberry Street, New York, 2008
View from the south. Local zoning regulations required the use of masonry on street-facing walls. The building was conceived as a Modernist massing clad in a 'contextual cloak' of brick. The design is intended to celebrate the possibilities of modern masonry technology, while acknowledging that the brick is panellised. The windows of these facades are punched, as is typical of older brick construction, but the openings are staggered to work with the repeat pattern of the rippled surface. The south wall has horizontal clerestory windows in the bedrooms and master bathrooms.

Prize in 2004 for becoming the first privately funded craft to enter the realm of space twice within a two-week period.

While there have been great advances in the science of building materials and components, there has been little progress in the means and methods of design and construction during the previous half century. However, developments in digital technology are now creating an impetus for creative intervention in this arena. One is the automating of fabrication, and another the streamlining of communication. As digital technology begins to filter into the everyday lexicon of the construction process, becoming the norm as opposed to the exception, it is initiating a sea change in the way buildings are planned, designed and built. At the centre of this process is building information modelling (BIM).

The benefits and details of BIM are well documented and beyond the scope of this discussion, which will focus on the use of BIM to promote communication and integrate the discrete knowledge of many disparate parties into a holistic process of planning, design and construction.

From Representation to Innovation

Architects' first use of three-dimensional modelling was as a tool to communicate with clients, by offering realistic renderings and 'walk-throughs' of their buildings. But these models were dumb; literally empty shells wrapped in texture maps. Today's building information model is embedded with real-time data that can be extracted for well-informed and timely decision making by many users throughout the life of the project. Clients, in addition to visualising what the building will look like, are empowered with marketing, financial, operations and life-cycle costing information. This knowledge, in turn, informs a richer dialogue and a more robust relationship between owners and the design team.

In working with consultants, BIM should be far more than a simple tool for 'clash control' (the identification of conflicts between building systems). Use of the model and associated analytic software (for example, structural and environmental analysis programs) can improve the quality of such systems by incorporating their parameters into the basis of the architectural design. When consultants (whose work typically lags behind the architect's by at least one phase) enter into the project, the systems design has already begun. The architectural team is educated and aware of the issues, and takes an active versus passive role in the further development of the systems and integration with the overall design. In addition, not only does the building information model need to serve as a repository for increasingly complex information during the design process, it also needs to be assembled in such a way that each consultant and fabricator or contractor can later extract the specific information they need. Information is only valuable if it is applicable, and a wealth of non-applicable information is actually confusing and a deterrent to using the model. It is therefore crucial that the information is edited and made available to each user based on their individual needs.

Parametric coordination of structure, brick coursing and window openings. The staggered pattern of the windows gave very little leeway to adjust the location of columns. In order to optimise panel efficiency, any local adjustments needed to be applied globally.

SHoP Architects, Rector Street Bridge, New York, 2007
Physical model of the Rector Street pedestrian bridge proposal. The original box-truss superstructure supports the walking surface, while the composite frame supports, and gives shape to, the enclosure. In the model, this exoskeleton is a digital print, producing an immediate translation from the virtual to the physical. In reality, the actual scale and material fabrication processes require intermediate steps such as moulds, templates, assembly of smaller components and so on.

Breaking from Tradition: A Team Approach

The precision with which building components are modelled forces architects to engage with fabricators and other trades early in the design process. Furthermore, when the fabrication process is automated, the technology becomes a shared platform of design and communication. The logistical barriers to working with subcontractors prior to letting a contract challenge the design-bid-build method. Certain trades may be engaged early in the design process, on the basis of partially complete documents, and their input can be utilised to finalise the documents, or they may be engaged independently of construction for 'design assist' services.

The BIM environment thus tends to encourage an integrated project delivery (IPD) approach in which parties to the project band together contractually in a unified, collaborative effort. In the traditional owner-architect-constructor model, the architect and the constructor have an adversarial relationship. The joint-venture nature of the IPD concept not only adds value to the design process by engaging a wider knowledge base at an earlier stage of the project; it also places parties on the same team and changes the liability landscape. Assuming properly crafted contractual relationships, risk-management activity shifts from reactive (finger-pointing after the fact) to proactive (where team members work together to target and prevent costly mistakes before they happen).

Prior to construction, the building information model becomes a way for the team to practise-build the project. In order to be effective, those who build the model need to be in the field in some capacity during construction, otherwise this knowledge is wasted. Those who have created the model have 'lived' in the building for months (or years), can anticipate potential conflicts, understand sequencing issues, and can recognise when a constructed element looks wrong or out of place. The later a conflict is discovered, the greater the potential cost in terms of delays and correction of the work. The trend towards outsourcing BIM overseas, to save on labour costs, is therefore counterproductive to the intent of the exercise.

Case Studies

Rector Street Bridge, New York, 2007

Immediately following the attacks of 11 September 2001, SHoP was asked by Battery Park City Authority (BPCA) to design a temporary pedestrian bridge that would reconnect Battery Park City with the rest of Lower Manhattan. The bridge was designed with a largely opaque enclosure to prevent it being used as a viewing platform for the World Trade Center site. Constructed in 2002, Rector Street Bridge was intended to be taken down in 2004, and has now outlived its intended use.

At the time of writing, several options for reconfiguration or relocation of a new bridge are under consideration. One proposal is to open up the bridge to views over the city and use materials more consistent with the goals of the proposed 9A Greenway. This new bridge would reuse the existing structure, thus there is a predetermined maximum design load. In addition to weight limit, performance criteria include passive thermal control (shelter from wind and resistance to heat gain), drainage and ice control (water and ice must not be shed to the road below), protection of the pedestrian walkway from rain and snow, lighting conditions (maximising daylight while reducing glare and providing optimum lighting at night) and air quality.

SHoP began by looking at two lightweight materials: ethylene tetrafluoroethylene (ETFE) supported by a glass composite skeleton (the original box-truss superstructure supports the pedestrian walkway and the enclosure, but a secondary structure is required to give form and provide more flexible points of attachment for the enclosure). ETFE is a high-strength plastic that can be used in pillow-like pneumatic systems for lightweight, transparent or translucent long-span enclosures.

For the design of the composite structure, SHoP worked with custom boatbuilders New England Boatworks, who were contracted by BPCA in a design-assist capacity. These fabricators brought to the project not only their wealth of knowledge and expertise in the design and construction of complex composite structures, but also their enthusiasm and unique viewpoint. It soon became apparent that both architects and fabricators shared a deep passion for structural and design optimisation, and utilised many of the same digital tools. SHoP also worked closely with ETFE cladding consultants Vector Foiltec, with SDK Structures for the engineering of the composite skeleton, and with Buro Happold as general structural and mechanical engineers for the project. In addition, students from the product architecture and engineering programme at Stevens Institute of Technology in Hoboken, New Jersey, used the bridge as a case study for daylighting analysis as part of their coursework. One such study focused on the potential of altering pattern densities on the ETFE to reduce glare and heat gain while maximising views and daylight.

Software compatibility was crucial to the bridge project. Custom macros were developed to extract information, such as dimensions and key point coordinates, and transfer it to other applications. This allowed all of the participants to use the best software for any particular task while keeping all models coordinated to avoid errors. (Many of the information transfer tools and protocols were developed by the students at Stevens Institute.)

In theory, automated fabrication holds the promise that variable components can be produced at close to the same cost as standardised components. For example, a laser cutter working from a digital file can cut many different shapes for the same cost as cutting the same number of identical shapes. The adoption of parametric modelling affords a similar economy to the design process. Rather than individually detailing each situational instance of a varying design, a single prototypical instance can be detailed that anticipates varying situations and governs the outcome. Indeed, the design of the guardrail for the approaches to the bridge took advantage of this methodology. The rail consisted of

INTEROPERABILITY

shop - rector street bridge - current information transfer

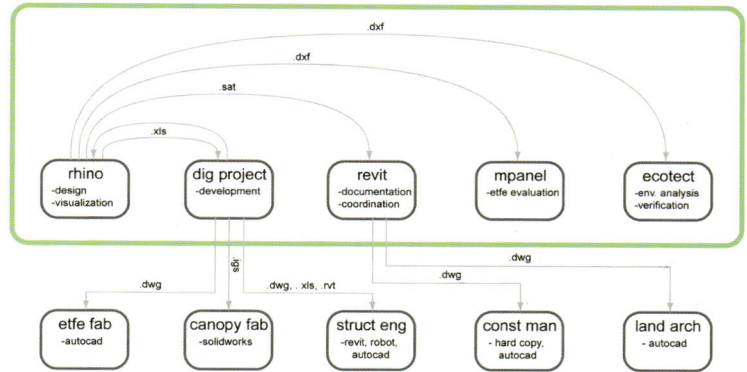

The design workflow for the bridge broken into three parts: canopy, structure and site. The early design (concept/schematic) was developed through traditional iterative explicit modelling (Rhino) and 2-D drawing (AutoCAD) in order to facilitate quick and varied design changes. Once the overall design approach was identified, the canopy model was rebuilt in a parametric solid modelling platform (Digital Project) to enable more robust and detailed model development. The structural model was built in a parametric BIM platform (Revit) for coordination and documentation with the canopy model. The site work remained in 2-D, but was referenced into the coordinated BIM model. The parametric canopy and structural models were the primary ones used throughout design development and preparation of the construction documents, although specific design exercises were carried out in other modelling platforms: Rhinoceros for explicit modelling and Autodesk Inventor for design studies and fabrication models. All documents, both internal and external, were coordinated through the building information model.

INCIDENT SOLAR RADIATION: SUMMER

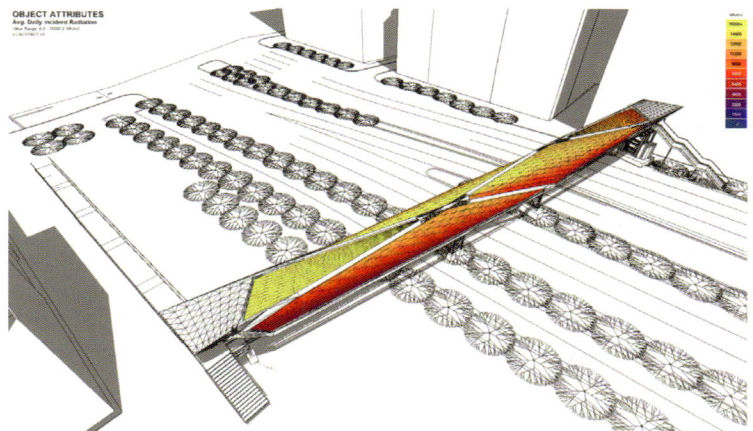

Simulation of the amount of solar radiation hitting the ETFE pillows throughout the summer on the bridge canopy, suggesting that the pattern should be densest on the portions coloured yellow (at the bridge roof) to reduce glare and heat gain. While seeming self-explanatory, the amount of local adjustment in the information model to modulate openings was critical to the performance of the canopy. Studies concerned the use of lighting and solar radiation analysis to generate potential pattern densities for the ETFE in order to reduce glare and heat gain, while maximising views and daylight. By varying the pressure in the ETFE pillows, printed patterns can be programmed to change throughout the day in response to changing lighting requirements.

punched and folded metal panels, the governing design for which was the same. However, the dimensions and geometry of the individual panels needed to vary to respond to changing conditions – such as the slope of a ramp versus a stair, and so on. Simplified representations of the panels were developed in the building information model. Automated routines then extracted key parameters from this boundary geometry and used them, in another software, to build detailed models suitable for automated fabrication.

The final design of the canopy structure was optimised for cost efficiency and structural and environmental performance. While form and aesthetic were always considered and shaped by the members of the design team, the specific end result grew out of the computational design process. Collaboration with specialist fabricators in the early stages of the design process allowed for a fluid dialogue between design objectives, fabrication constraints and the potential of the materials.

SHoP Architects, 290 Mulberry Street, New York, 2008
The master (positive) from which form liners (negatives) are made represents a significant part of the panel fabrication cost. Once the general concept was determined, the design was driven in large part by the goal of optimising the efficiency of the master. The size and shape of individual panels were driven by aesthetic, structural, shipping and installation concerns.

290 Mulberry Street, New York, 2008
This 13-storey residential condominium building, developed by Cardinal Investments, is located in New York City's Nolita district, across the street from the historic Puck Building, one of the city's most recognised and distinctive masonry buildings. Local contextual zoning regulations required a predominantly masonry facade. The design concept thus grew out of a desire to explore and celebrate the possibilities of modern masonry construction technology. The rippled brick facade treatment acknowledges the fact that the brick is panellised – the ornamental motif of the ripple would have been difficult, if not impossible, to lay by hand, and the panels clearly read as a curtain wall as opposed to attempting to mimic load-bearing or infill construction.

Prior to this project, SHoP had used parametric modelling only to resolve complex geometries or for digital fabrication purposes. Base buildings were still drawn in 2-D using traditional methods. 290 Mulberry Street therefore became the pilot project for the implementation of BIM as the office standard, and was used to understand the capabilities of the software, train staff and develop best practices including consultant interface and interoperability with other software tools. This interoperability is of paramount importance in order to effectively use the best tools for any given task. For the Mulberry Street building, the complex geometry of the facade was developed and detailed outside of the building information model, and boundary geometry imported for reference and coordination purposes only.

Initial conversations with brick-panel fabricators informed the general design of the facade. However, parameters vary between fabricators (for example, weight of concrete backing, type and coverage of reinforcement, preferred method of attachment and so on). The parametric model of the facade needed to be precise enough to account for very small tolerances in the brick coursing, and flexible enough to allow for modification as the panels were fully detailed. Scripts were developed in-house to control such factors as variation of brick dimension, coursing and minimum coverage of the bricks as they 'step' in the rippled pattern, size and location of panel joints and windows, and location of columns. Other considerations were size and weight of panels for transportation and installation, structural loading and cost. During the 'shop drawing' stage of the panel design, the fabricator (Saramac) built their own digital model, and this was repeatedly checked against SHoP's own model for inconsistencies.

The north and west facades of the building are brick curtain walls – precast-concrete panels with bricks embedded in the surface. They are produced in a factory, shipped to the site and installed with a crane. Panellised brick can offer higher quality control and lower labour costs than on-site cavity-wall construction. Panels are fabricated using a rubber form liner into which the bricks are set, steel reinforcing is placed and concrete poured. The master mould from which all the form liners are produced is in essence a composite 'positive' of the entire curtain wall, and was milled directly from SHoP's digital files. Because the cost of this component is high in relation to the overall facade cost, it was desirable to generate the greatest variety of panel shapes from

Panel fabrication. Bricks are set face down in rubber liners, reinforcement and embeds are placed and concrete is poured. For large runs, multiple liners can be cast from the master for the simultaneous production of identical panel types.

A typical panel cures in the yard prior to shipment. The back-up concrete is tinted to the desired colour of the mortar joints. The architects chose to expose this concrete at the window returns as well. Three typical panels come together to form a window opening, and matching concrete sills were cast separately and placed in the field.

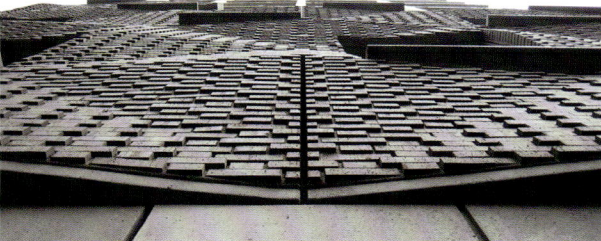

Completed and installed panel details. The joints between panels are integrated with the global design of the facade and panel order. Panels at the south wall, residential entrance and store front are cast from the tinted grey concrete that is used as the mortar material on the brick panels.

the least amount of master mould and form liner. While the panel fabrication was competitively bid, the master mould and form liner fabrication was specified as a sole source contract (through Architectural Polymers) allowing SHoP to secure this trade input and create mock-ups early in the design process.

As the Mulberry Street building is a speculative development project, the building information model also provided key data extraction for a variety of team members, including marketing and financial planning. Live links between the model and various zoning and code calculation schedules and legal documents also drastically reduced the design time required to keep this information current and error-free. As BIM technology becomes mainstream, it will no doubt in most sectors be utilised to cut costs while maintaining minimum design standards. For the rest of us, the hope is that by working more smartly we will be able to remain competitive throughout changing market cycles while devoting more of our time and effort to innovative design and improved construction quality.

BIM: Building Information MANAGEMENT

One can argue that there is no such thing as a bad decision, only an uninformed one: that is, in a world in which all information is known, one would always make the 'right' decision. Fluid communication is critical to information transfer. Getting information to people who need it for decision making lessens the probability of uninformed, 'bad' decisions. Transferring this information quickly reduces project timelines (people wait for information in order to make informed decisions, or they make uninformed decisions that result in redesign and construction delays) and waste (from construction errors and over-estimation of materials). As information is shared among the parties working on a project, the knowledge base of the group grows exponentially and contributes to a higher-quality end result.

SHoP has recently launched a construction management firm, SHoP Construction Services, to provide construction advisory and construct-assist services, and also to serve as a link for the architecture firm to the knowledge pool of the construction trades. The project management strategy of the new company focuses on the building information model for information sharing. Through IPD (various joint-venture models) it is securing the input of allied trades early in the design process to build robust, fully coordinated models. These models will be kept 'live' during the construction process to reflect as-built conditions and detect and resolve discrepancies. A future goal of the company is to develop seamless tie-ins between the building information model and automated facility management systems, providing an invaluable tool for building owners and allowing the practice to better integrate life-cycle costing and building management information into the future planning, design and construction of new buildings. ∆

C² Building, Fashion Institute of Technology

New York, 2009–

SHoP Architects

By SHoP Architects

The proposed building is highlighted by a multilayered glass and metal facade, within which are nested the primary circulation, review and exhibition spaces that connect the building's design studios with the sky-lit student quad on the fifth floor. Just as a loom builds form and structure simultaneously, this new kind of building allows structural systems, environmental technologies and visual permeability to be interwoven in new construction concepts.

The C^2 Building is centred on the creation of a new, urban, vertical Student Life Hall that is open to students all year round. Two 49-metre (160-foot) long steel trusses span between the east and west cores of the building creating an uninterrupted open floor-plan for the hall. This design strategy establishes a highly flexible three-storey space supporting diverse student activities such as casual gatherings, music performances, lectures and fashion shows.

Flexibility, communication and leading-edge technology are what underpin the C^2 Building, which it is hoped will be a unique and inspiring example of future design, creating a transformative environment for students, faculty and alumni as well as the people of New York.

The new 10-storey, 9,132-square-metre (98,300-square-foot) C^2 academic building.

As the main establishment for fashion and design education in the US, the Fashion Institute of Technology is unique in that it is located directly in the heart of the fashion industry it teaches. However, the school lacks a clear sense of place within the city. One of the primary goals of SHoP's design for the new C^2 academic facility was therefore to create an iconic building that would form a lasting identity for the school, and one that would also functionally link existing academic spaces within the adjacent campus buildings with new classrooms, faculty and administration offices, and a sunlit student hall for gatherings and events.

The new building is highlighted by a multilayered glass and metal facade, within which are nested the primary circulation and the review and exhibition spaces, connecting the design studios with the skylit student quad on the fifth floor. An express escalator takes students directly from the street lobby to this floor, which is also the point at which adjacent buildings on the campus connect.

The C^2 addition will be a LEED-certified project and, in a state-sponsored initiative, the upper portion of the south-facing facade of the atrium will house a new and experimental dynamic solar curtain-wall system to reduce heat gain and control glare.

Just as a loom builds form and structure simultaneously, this new building type allows the simultaneous interweaving and construction of its structural systems, environmental technologies and visual permeability.

The structure consists of four main parts: the lower-level framing system that houses classrooms and laboratories; the vertical trusses that contain the building's cores and supports; the long-span steel trusses that bridge over the column-free fifth-floor student lounge; and the cable facade that encloses the primary building circulation.

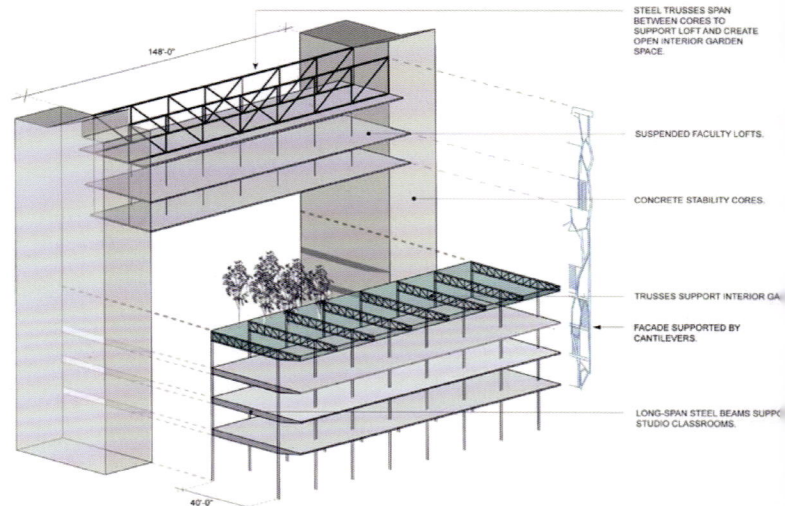

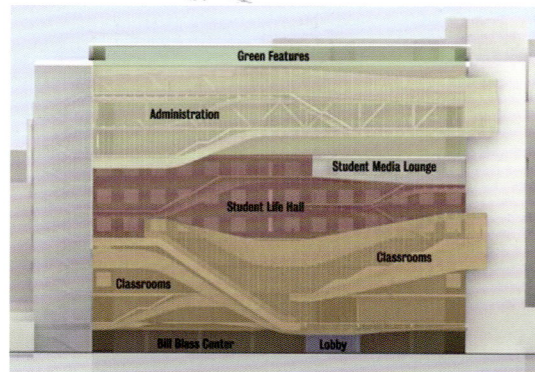

STEEL TRUSSES SPAN BETWEEN CORES TO SUPPORT LOFT AND CREATE OPEN INTERIOR GARDEN SPACE

SUSPENDED FACULTY LOFTS.

CONCRETE STABILITY CORES.

TRUSSES SUPPORT INTERIOR GA

FACADE SUPPORTED BY CANTILEVERS.

LONG-SPAN STEEL BEAMS SUPP(STUDIO CLASSROOMS.

Green Features

Administration

Student Media Lounge

Student Life Hall

Classrooms

Classrooms

Bill Blass Center   Lobby

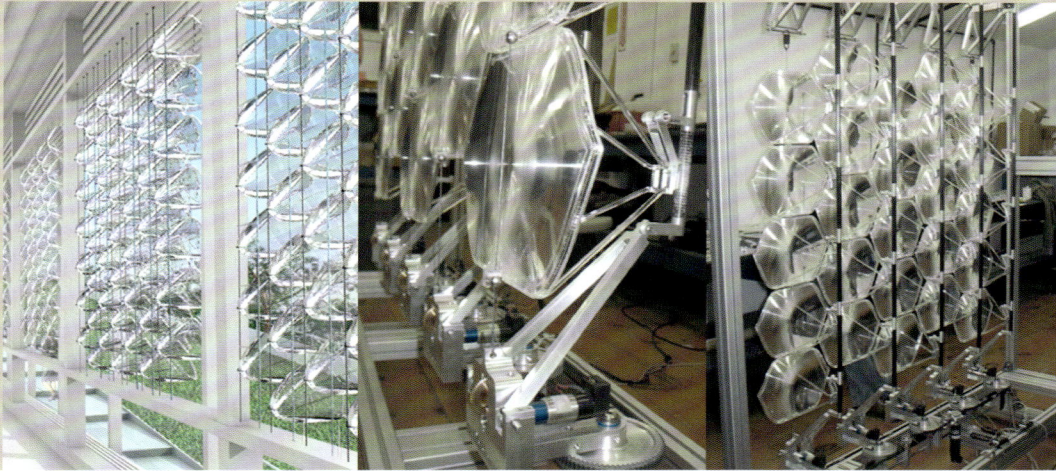

IC Solar Facade is a next-generation photovoltaic facade system that produces three to four times the amount of energy of the best photovoltaic technology currently available. Its electrical efficiencies are the result of the innovative use of Fresnel lenses in conjunction with advanced solar-tracking algorithms. In addition to its noteworthy power production, the system also reduces building cooling loads, improves daylighting efficiencies and generates hot water that can be used throughout the building.

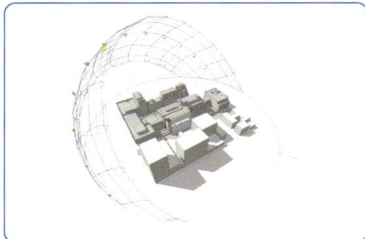

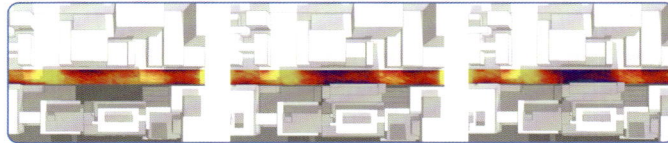

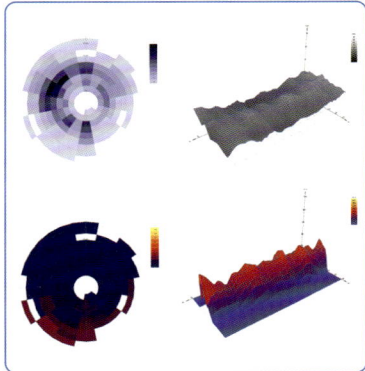

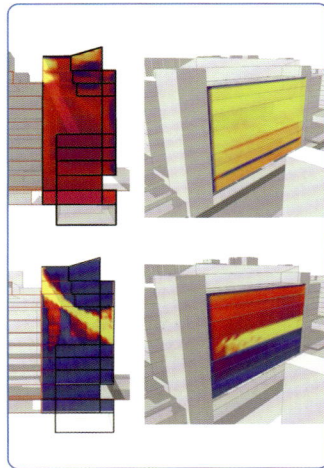

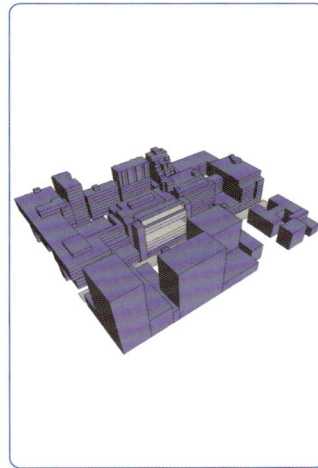

Design research for the project focused on replicating, analysing and scrutinising the predicted environmental performance of the C^2 Building. This work concentrated on three specific topics: site and climate analysis, daylighting analysis and thermal analysis. The resultant models provided the means by which designing through iterative analysis could be carried out, and laid the foundations for a highly intelligent parametric DNA from which the building could be designed. This model formed the basis for re-creating these efforts in advanced software packages so that an Energy Cost Budget model could be developed in accordance with ASHRAE 90.1 standards. ⌂

Automated Assessment of

Building information modelling (BIM) is a powerful tool for clients and architects alike, particularly when clients have ongoing complex programmatic requirements. **Chuck Eastman** describes how with his team* at the AEC Integration Laboratory at the College of Architecture at the Georgia Institute of Technology he was commissioned by the US federal government's General Service Administration (GSA) to automate the design guidelines for all US courthouses in such a way that preliminary designs by architects could be assessed and checked against specific criteria.

Early concept designs are hugely important in determining the eventual success and impact of a project. The work of Louis Kahn, Alvar Aalto and Frank Gehry, among others, has shown how their initial concept sketches eventually determined the final project, in terms of creativity, costs, support for the building's functions, visual impact and other more general factors. Further development and detailing later on in the process can refine and elaborate a good early concept, but can only partially ameliorate a bad one.

Until now, concept design has been a largely mental exercise of generating various spatial concepts and assessing them intuitively, based on the designer's knowledge and accumulated expertise. Reliance on such expertise is perhaps one reason why architectural success has traditionally come only to those with decades of experience who are able to bring to bear the wisdom required to assess and select design concepts worthy of being fully developed.

While the importance of early concept design has long been recognised, digital tools to support this stage of the design process have been sparse and largely unsuccessful. Perhaps the only significant exceptions have been 3-D sketching tools such as Sketch-up and form-Z, which allow the development of concepts in three dimensions rather than two. However, while such tools substitute a 3-D sketch for a 2-D one, they do not help designers when it comes to augmenting expertise.

The construction industry is undergoing a revolution in terms of data representation. Twenty-five years after the transition began in the aerospace and manufacturing sectors, architecture and construction are now following suit and relying on digital parametric models of the designed product.[1] Another capability developed in manufacturing was interoperability between applications using ISO-STEP exchange standards; the equivalent in construction are the Industry Foundation Classes (IFC).[2] Thanks to the groundwork carried out by these other industries, the transition of architecture, engineering and construction

to what has been coined building information modelling (BIM) will probably be much faster as they select from, adapt and add to the methods already developed. More than half of all architectural firms in the US now claim to be using BIM.[3]

Members of the AEC Integration Laboratory at the Georgia Institute of Technology's College of Architecture have been working to advance digital design practices within the process framework defined by the US federal government's General Service Administration (GSA), which is the real-estate management arm of the Department of Commerce. The GSA is one of the main facilitators of the building industry's move to BIM and is responsible for the design and construction of all US courthouses.[4] A courthouse provides the space necessary to carry out the functions of the US judicial system, and the design thus entails complex issues regarding circulation and security. The programmatic requirements and best practices for courthouses are spelt out in the *US Courts Design Guide* (2007), a 400-page document that outlines the spatial and environmental requirements, circulation, communications, security and other factors particular to courthouse design.[5] The AEC Integration Laboratory was commissioned to begin automating these aspects of the design guide and to work on the design reviews required for courthouse planning.

The GSA has a very well-defined design process for public buildings, including courthouses, that is spelt out in its *P-100 Facilities Standards for the Public Buildings Service* (2005) design guide.[6] This sets out the process, deliverables, reviews and iteration cycles required to execute a new design and construction project, and a number of these steps are now being modified to adopt BIM-enabled processes. The guide outlines the planning and feasibility steps for a project prior to the contract being awarded to an architectural design firm. These include the development of a Housing Plan that identifies all of the space for which federal funding is required. This initial plan is then refined and expanded to generate a cost estimate that forms the basis for the application for congressional authorisation and funding. Once funding is approved, the architects and engineers for the project are selected.

Upon selection, the architectural firm gains information from many sources, including the *US Courts Design Guide* and also CourtsWeb, a case-based website of courthouse design issues.[7] It is from these, and through discussions with the relevant court and GSA staff, that the initial design concepts are generated.

Early Concept Designs

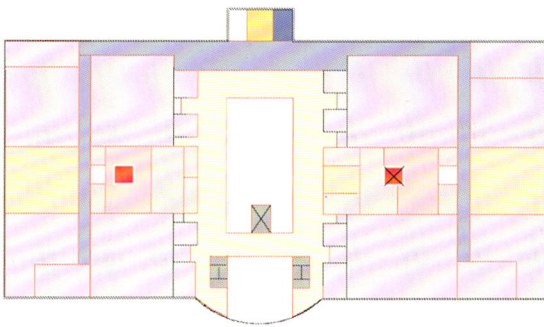

Two floors of a test building model for early concept design. The models consist of 3-D departmental-level spaces laid out on floor slabs without interior separating walls, but with exterior walls for each floor level.

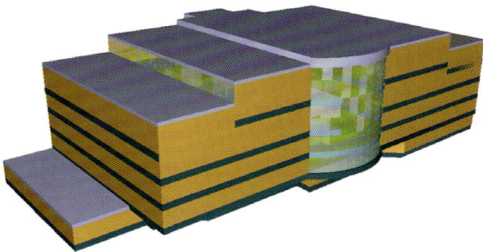

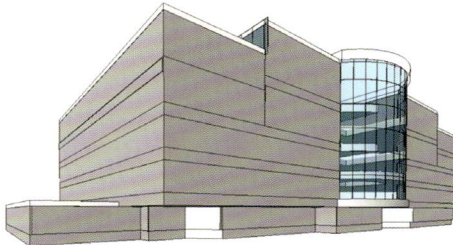

Massing studies generated from the early concept design model shown above.

The *P-100 Facilities Standards for the Public Buildings Service* design guide defines the content of the Preliminary Concept Designs that are to be submitted by the architects-engineers for review. These provide a narrative overview of the site – its setting, history and context – and also outline general considerations regarding visual style, site characteristics such as local density and proposed building height, and materials. Along with the general context, the architectural firm generates multiple spatial concepts. The GSA requires at least three, and more are usually generated through refinements and iterations.

Preliminary Concept Designs traditionally consisted of a site plan that showed the enclosed and outside parking and floor plans of a building, with elevators, stairways and mechanical spaces, so that the gross and net areas of each floor could be assessed. Floor-to-floor and ceiling heights of the spaces were also required, providing 3-D information that was presented as massing studies and renderings of the design concept. In this pre-BIM world, these early designs were presented in paper format, as floor plans, diagrams and renderings. The GSA appoints a design review board and carries out background studies of each submitted design concept, in terms of its relation to the space programme, codes and standards (including fire regulations and access) and compliance with the *US Courts Design Guide*. It also generates preliminary cost and energy use estimates to determine whether a proposal is within the scope of the government's budget allocation and energy-efficiency targets. All this was previously done by hand, by GSA staff or consultants, and involved days of tabulation.

More recently, architects have begun to submit Preliminary Concept Designs in the form of 3-D building models, which means their proposals can be partially assessed automatically. The concept design can now be generated using any of the GSA-approved BIM design tools. Currently these include Revit, Bentley Architecture and ArchiCAD, but others such as Digital Project, Vectorworks and Allplan are also being considered. However, they must include the following, as set out in the 'GSA Preliminary Concept Design BIM Guide' (2008)[8] prepared by the AEC Laboratory at Georgia Tech:

- floor slabs defined with target thickness and floor-to-floor distances – also used for the roof;
- a composition of 3-D space objects, carefully named, on each floor slab, with height designating ceiling height;
- exterior walls with no construction, but with per cent glazing and R-values;
- building placement on the site, with orientation and above- and below-grade designation.

The above information is the minimum required to define a Preliminary Design Concept, but is detailed enough to generate meaningful assessments. Each of the GSA-approved BIM design tools supports the file export of a design model in IFC format, the international standard neutral representation of building model data. The file is read in an application suite developed with Solibri Model Checker[9] as a platform that supports the following automated assessments based on data read from the IFC file:

1 Spatial validation of the layout, comparing target counts and areas of the courthouse project space programme with those of the proposed concept design.
2 Circulation analysis of the layout, based on the courthouse-specific criteria of the *US Courts Design Guide*.
3 A preliminary energy assessment, using the Energy-Plus analysis tool.
4 A preliminary cost estimate, using the PACES cost-estimating system.

The results provide a uniform set of assessments, guaranteeing that the same assumptions and criteria are used for the different variations and iterations of the same concept design and, over time, across multiple projects.

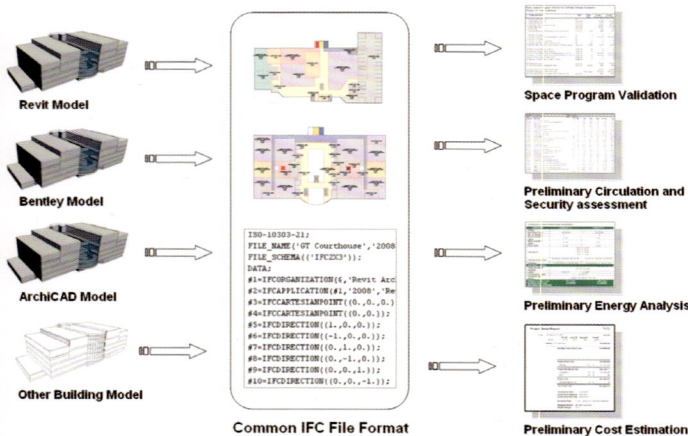

The general configuration of the GSA Preliminary Concept Design assessment tool. Most of the BIM design tools can generate Preliminary Concept Design models and export them to IFC. The IFC model can be automatically interpreted to support four assessments: for space programme validation, circulation and security analysis, preliminary energy analysis and cost estimation.

As a general syntax- and content-checking application, the Preliminary Concept Design prechecking review tool assesses whether the submitted building model has the correct elements, naming conventions, properties and other structures needed for full assessment. It returns diagnostic reports of the submitted model, and determines if and what corrections are needed. Prechecking in this way ensures that incorrect building models do not lead to meaningless analyses.

Space Names for Preliminary Concept Design Assessment

The spaces within a building are named differently according to application needs and life-cycle stage. Each of the supported applications has different naming conventions: rentable space names are different to those from the *US Courts Design Guide*, and those used for cost estimation are different to those used for energy analysis. While the long-term objective is to develop a master set of space names for a building type that covers all uses, it could be years before such an undertaking is agreed on. The AEC Integration Laboratory has thus developed a name-mapping method that automatically maps space names for their different uses. The master space name-set is categorised into elementary and aggregation space names. Departmental spaces and individual base spaces are often mixed in a concept design, and the Preliminary Concept Design review tool can accept such mixtures. Where departmental spaces are shown, the percentages of space allocation within a department, based on the building space programme, are used to estimate the areas of the individual spaces.

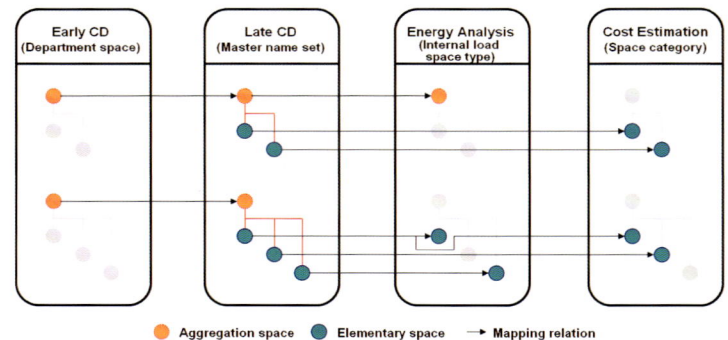

Mapping relation of space names. Names are mapped to their most disaggregated base spaces, and then re-aggregated into the classes needed for departments, circulation and security analysis, energy analysis and cost–estimating applications.

Space Programme Validation

The Space Validation application is an adaptation of the Late Concept Design Spatial Validation BIM application developed by Solibri for GSA. The application applies GSA-specific rules for area calculation and reconciles a design with the Congressionally authorised space programme. For early concept design models, the AEC Integration Laboratory developed a version of the space validation application to deal with the given level of detail of the model. This compares multiple alternative layouts to the target space requirements of the space programme and includes the efficiency and adequacy parameters traditionally used by the GSA to compare alternatives.

The review presents the given building model's space programme in an architect-friendly form; for example, number of spaces, gross area, usable area, building efficiency and so on. The figure here shows one of multiple space reports generated.

Early Space Program Review for Concept Design Evaluation
Project: GT Test Courthouse

#	Design parameter	Type	Target Value	Concept 1 Actual Value	Concept 2 Actual Value
1	Number of Building Floors	EA		6	6
2	Total building gross area	Area (nsf)		197,269	201,005
3	Inside parking area	Area (nsf)		10,319	10,380
4	Total gross minus inside parking area	Area (nsf)		186,950	190,625
5	Total usable area	Area (usf)		159,317	161,100
6	Atrium area	Area (nsf)		622	622
7	Building Efficiency (USF/Total gross minus parking area)	Ratio (%)	67%	85%	82%
8	Number of Courtrooms	EA	9	9	9
9	Number of Special Proceedings/Appeals Courtrooms	EA	0	Not found	Not found
10	Number of Chambers	EA	11	11	11
11	Number of Inside Parking Spaces	EA	24	22	24
12	Number of Elevator Spaces on the 1st Floor	EA	TBD	6	6
13	Elevator Ratio (Total Gross Area / Number of Elevator Spaces)	Area (nsf)	25,000	32,878	32,878
14	Floor to Floor Height for Courtroom	Height (ft)	20	20	20
15	Maximum Ceiling Height of Courtroom	Height (ft)	16	14	16
16	Floor to Floor Height for Sp. Proceedings/Appeals Courtroom	Height (ft)	-	Not found	Not found
17	Maximum Ceiling Height for Sp. Proceedings/Appeals Courtroom	Height (ft)	18	Not found	Not found
18	Floor to Floor Height for Office Space	Height (ft)	14	14	14
19	Maximum Ceiling Height of Judges Chamber	Height (ft)	10	10	10
20	Building Skin Area	Area (nsf)		99,579	100,422
21	Total Gross Area to Building Skin Area	Ratio (%)	45-55%	50%	49%
22	Main Entrance's floor level (Ground Level)			Level 02	Level 02
23	USMS Administrative Office's floor level		2nd or upper	Not found	Not found
24	Gross Area of Prisoner Circulation and Holding Cell Area	Area (nsf)		14,902	14,902

Example space programme validation, assessing two candidate designs against the a priori space programme. Data from multiple alternative models are compared with the requirements in a single page.

Preliminary Circulation and Security Assessment

The *US Courts Design Guide* has many criteria related to circulation and security. A courthouse generally has three circulation systems that must not intersect or overlap. One is for the public, another for the judges, jury and court staff, and the third is for defendants and US marshals. These criteria are a major determinant of the form of a courthouse space plan. For the *US Courts Design Guide*, the research team identified 216 different statements defining circulation issues to be checked. These were in the form 'Each appellate court shall have …', meaning that it applies to all instances of appellate courtrooms. Some of the statements applied to 'All courtrooms shall …', which means to all types of courtroom. Since a courthouse can often have more than 10 courtrooms of various types, the 216 rules are in this case multiplied many times.

In the Preliminary Concept Design, only the department-level spaces are defined. In most cases, walls inside walls are not defined, thus exact circulation paths cannot be assessed and only a subset of the circulation rules can be applied.

The specific space types have one of three security types: public, restricted, and secure (for defendants and US marshals). Among the 216 circulation rules in the *US Courts Design Guide*, 43 per cent involve accessibility between two spaces within the same security zone. These can be checked simply by identifying their existence in the same zone. In order to check the containment of spaces in a zone, the spaces adjacent and having the same security level are structured as a logical set. The test is almost instantaneous.

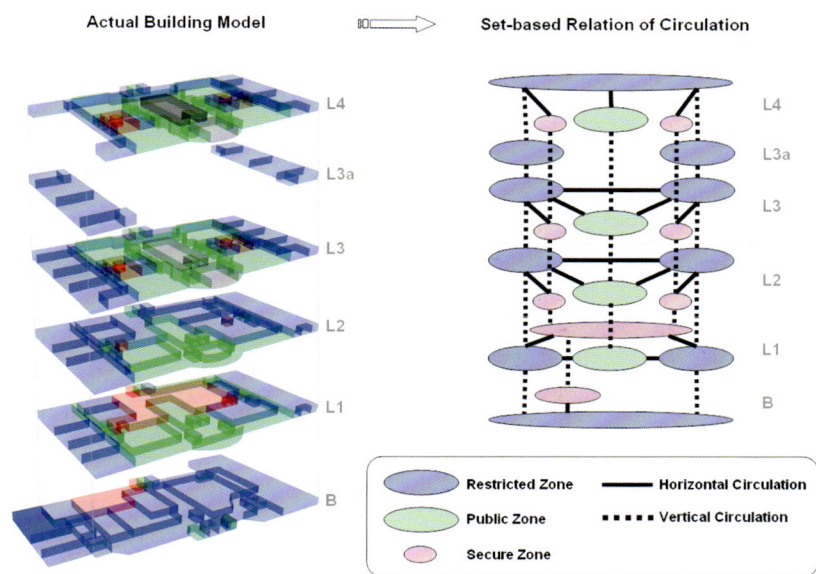

Actual Building Model ➡ **Set-based Relation of Circulation**

Restricted Zone — Horizontal Circulation
Public Zone ▪▪▪▪ Vertical Circulation
Secure Zone

In the abstraction used for Preliminary Concept Design circulation analysis, spaces are grouped into sets that are adjacent and have the same security. The connectivity of these zones is represented as solid edges, vertical access as dotted edges. If the specific circulation rule requires accessibility within a security zone and floor, then the vertical connections are disregarded.

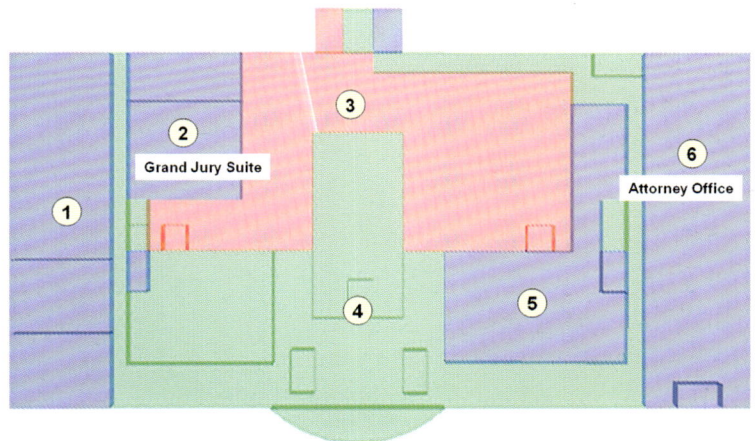

An example circulation-checking rule for courthouses is that the 'Attorney Office' should be accessible to the 'Grand Jury Suite' through a restricted zone. In the test model here, there are six zones according to adjacency and security level. But, even though the Attorney Office and the Grand Jury Suite have the same security level (restricted zone), public zone number 4 is placed in between the two target spaces, violating the circulation rule.

Preliminary Energy Analysis

An early concept design has features that significantly determine energy-use ranges. These include building orientation, the building shell's external materials, floor-by-floor footprint, insulation and the inclusion of atria, courtyards and skylights. At this stage designers are interested in a proposed building's heating and cooling loads over the year, required to condition the space within a particular comfort zone. The intention is to assess the impacts of these and other features that may significantly affect energy usage, and to facilitate design decisions leading to better energy performance.

In order to run the EnergyPlus analysis tool with this limited information, default values are provided based on typical values by building type. It is assumed that the courthouse will be in a city or a town, thus the solar distribution is set to that for an urban setting. At this stage the mechanical system is 'idealised' to supply the necessary heating and cooling. Values for internal heat gains such as occupant density, lighting and equipment loads are derived from the spaces in each of the building's thermal zones, such as the courtroom, judge's chambers and clerks' offices.

Building zones are an important aspect of an energy model. For preliminary energy analysis, a perimeter and core thermal modelling approach is used. Preliminary reporting samples are shown here.

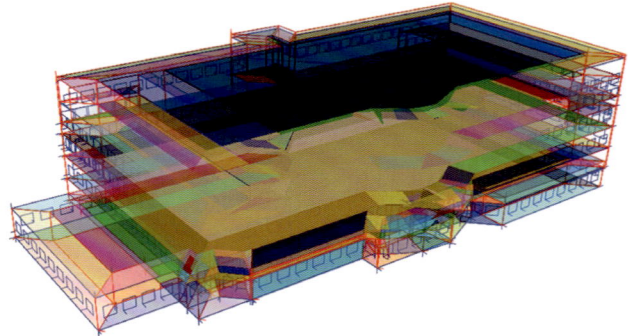

Example of the current method of automatic thermal zone generation, based on floor-by-floor perimeter zoning.

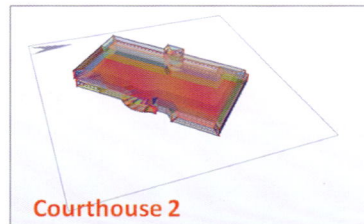

Example feedback from the energy analysis module. The effect of small building rotations is reported to assess orientation sensitivity and month-by-month energy usage for heating and cooling.

Courthouse 1

Date/Time	test1 [30]	test1 [15]	test1 [0]	test1 [-15]	test1 [-30]
January	303,152,050,799.75	303,949,350,508.26	303,149,390,498.23	301,187,294,058.48	299,308,686,718.31
February	172,304,249,143.02	171,905,272,436.69	170,229,703,861.53	168,717,570,060.43	167,399,165,289.12
March	195,999,546,815.09	194,441,497,850.32	192,537,699,095.94	192,410,348,927.48	191,902,676,487.36
April	112,630,023,894.26	110,801,671,042.42	109,725,289,295.62	110,277,806,783.76	111,253,843,437.67
May	78,344,549,242.66	77,860,280,815.64	77,774,090,390.60	78,282,840,907.41	79,294,460,203.82
June	71,957,926,319.69	73,369,948,172.53	74,589,953,666.62	74,911,218,573.71	75,162,581,191.34
July	73,067,000,840.50	75,474,601,283.84	77,608,475,350.35	78,369,455,065.79	78,454,599,066.50
August	66,605,119,894.51	66,994,085,946.66	67,725,538,624.74	68,979,922,738.35	70,076,717,302.11
September	69,419,145,580.60	68,481,736,923.41	68,023,131,079.01	68,807,689,308.27	70,110,390,801.96
October	90,279,395,166.08	88,728,019,614.70	87,171,298,292.53	86,596,725,836.57	85,881,605,448.20
November	176,752,662,352.58	176,947,452,483.45	175,659,605,666.82	173,893,846,250.91	172,323,936,167.50
December	272,043,514,420.57	273,118,696,380.44	272,892,968,756.63	271,595,927,792.71	269,716,354,976.49
TOTAL (J)	1,682,555,184,469.30	1,682,072,613,458.36	1,677,087,144,578.61	1,674,030,646,303.89	1,670,885,017,090.37

Courthouse 2

Date/Time	court [30]	court [15]	court [-15]	court [-30]
January	321,836,663,521.24	323,159,203,147.96	322,718,657,566.26	321,060,864,504.03
February	187,074,505,500.22	187,290,328,751.55	185,105,909,945.75	183,468,189,130.53
March	214,148,515,867.80	212,304,363,008.73	210,108,671,586.46	208,977,045,470.94
April	129,084,719,257.15	127,321,733,257.42	126,617,315,220.25	127,197,822,093.60
May	95,004,422,379.26	95,071,428,366.32	95,977,070,965.10	95,896,018,087.73
June	87,726,318,298.68	89,959,947,660.30	92,893,789,053.43	92,707,488,367.71
July	88,070,381,288.60	91,733,827,684.35	96,977,255,771.51	97,142,496,503.76
August	79,416,253,797.73	80,675,256,196.67	84,981,155,436.07	86,497,375,849.20
September	81,368,207,097.07	80,478,452,223.86	82,963,503,812.27	84,819,579,286.76
October	106,620,824,135.78	104,769,735,284.12	103,250,157,866.11	103,054,718,374.12
November	191,386,082,216.58	192,067,851,784.73	190,966,974,316.49	189,399,481,344.84
December	289,777,835,226.57	291,351,893,448.16	291,787,970,396.32	290,340,757,615.74
TOTAL (J)	1,871,514,728,586.67	1,876,184,050,814.17	1,884,048,431,936.01	1,880,561,836,628.94

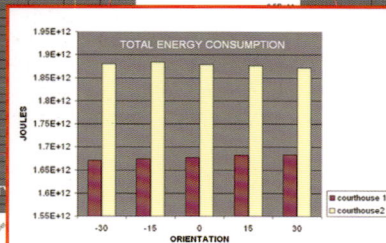

Preliminary Cost Estimate

Similar to the energy analysis, the intention of a preliminary cost estimate is to determine the effect of particular features of the design, and to gain insight into the value and potential cost of specific design concepts. The cost-estimation module is dependent upon two main components: building model-based data, and cost-driven data. Data from the building model includes all IFC-related information, such as mapped space names and their associated attributes, floor, roof and external wall areas, material properties and so on.

Cost-driven data includes defining cost types for different building types and spaces (for example, the costs for courtroom spaces are different to those for circulation spaces), and cost-calculation methods. The AEC Integration Laboratory uses Earth Tech's Parametric Cost Engineering System (PACES) software platform for this purpose.[10] Designers can specify the location of the building by entering the city and state, as well as the expected date and duration of construction, and fees and other cost data such as inflation, labour and interest rates. All values related to functional space areas, building structure and typical materials are mapped into expected quantities which are then priced according to assumptions regarding cost types, including labour, inflation and local availability. The assumed construction types and materials at the Preliminary Concept Design stage can then be tracked to see how the expected material quantities and costs vary as the design is detailed, providing a means of value engineering as design development progresses.

PROJECT DATA	
Project Name	Courthouse1
ADDRESS DATA	
City	Jefferson
State	Missouri
Country	USA
BUILDING DATA	
Facility Name	Jefferson City Courthouse
Model Name	Midrise Courthouse
Number of Floors above grade	6 floors
Number of Floors below grade	0 floors
Building Gross Area	197269 sq ft
BUILDING SHELL DATA	
Footprint	37303 sq ft
Perimeter	837 ft
Exterior Wall Area	49629 sq ft
Roof Area	37303 sq ft
Floor to Floor Height Above Grade	20 ft
Floor to Floor Height Below Grade	0 ft
Floor to Ceiling Height Above Grade	15 ft
Floor to Ceiling Height Below Grade	0 ft
FUNCTIONAL SPACES DATA	
Building Support Area	49406 sq ft
GSA 00/100 Open/Closed Office	5219 sq ft
GSA Conference/Classroom	759 sq ft
GSA Courtroom Low Rise	12989 sq ft
GSA Enhanced Office Low Rise	73478 sq ft
GSA Holding Cell/Detention Low Rise	7625 sq ft
GSA Joint Use and Retail	4729 sq ft
GSA Judicial Chambers Low Rise	11309 sq ft
GSA Judicial Hearing Room	2795 sq ft
GSA Library	4679 sq ft
GSA Underground Parking	20486 sq ft
Interior Loading Dock w/Exterior Canopy	1329 sq ft
Public Restrooms (Small)	2466 sq ft

Example of building model-based data extracted for use in cost estimation. The cost estimation is based on square footage and surface types and areas.

The Current Status of the Assessment Tool

Modules for the space programme review and the circulation and security review are operational and provide reports in little more than a minute for a five-storey courthouse. Most of the running time of this module is used in capturing images that will be embedded in the reporting document. The preliminary energy analysis module has been successfully integrated with the EnergyPlus simulation engine. The Preliminary Concept Design review tool has been provided with a particularly simple-to-use user interface that requires minimal operations in order to run the energy simulation. The Georgia Tech research team has begun the process of integrating the Preliminary Concept Design review tool into the PACES database. Once completed, this will require minimal or no input from the user to produce cost estimations representative of the analysed design phase.

During its development, the Preliminary Concept Design review tool has been operated solely by the AEC Laboratory team. However, the goal is to provide a tool that can sit on any designer's desktop, making the review process an integral part of the multiple iterations during the development of the preliminary design proposals. This also opens up the possibility for designers to develop their own sets of rules to be checked, which will allow experienced designers to pass on their expertise to those with less experience and make explicit the application of their design intentions for building projects.

Conclusion

The AEC Laboratory's work demonstrates the value of BIM for early concept design. The metrics applied provide quick feedback while designers are undertaking form/programme synthesis. These kinds of tools will not hinder the realisation of buildings with innovative visual character, but rather allow creative design to be better informed and debated. Such tools are also expected to allow younger designers to gain invaluable experience more quickly using virtual architecture assessments, and to facilitate the more rapid emergence of new ideas in practice. ∆

The work presented here was funded by the GSA Office of the Chief Architect. The authors wish to thank Calvin Kam, Fred Miller, Peggy Ho and Charles Matta for their support, and acknowledge the technical support of Pasi Passiala of Solibri.

Notes
*Project team at the AEC Integration Laboratory, the College of Architecture at the Georgia Institute of Technology: Jin Kook-Lee, Hugo Sheward, Paola Sanguinetti, Yeon-suk Jeong, Jaemin Lee and Sherif Abdelmohsen.
1. 'From Blueprint to Database', *Economist Technology Quarterly*, 7 June 2008, pp 18–20.
2. Industry Foundation Classes – IFC2x Edition 3, 2006, http://www.iai-international.org.
3. B Gilligan and J Kunz, 'VDC Use in 2007: Significant Value, Dramatic Growth, and Apparent Business Opportunity', *CIFE Technical Report #TR171*, Stanford University, December 2007.
4. See www.gsa.gov/bim, accessed 12 July 2008.
5. Judicial Conference of the US, *US Courts Design Guide*, Washington DC, 2007.
6. US General Services Administration, Office of the Chief Architect, *P-100 Facility Standards for the Public Buildings Service*, Washington DC, March 2005.
7. See http://www.publicarchitecture.gatech.edu/Research/project/courtsweb.htm.
8. 'GSA Preliminary Concept Design BIM Guide', Draft, College of Architecture, Georgia Institute of Technology, 2008.
9. See http://www.solibri.com, accessed 18 July 2008.
10. See http://talpart.earthtech.com/PACES.htm.

Cellophane House

New York, 2008

KieranTimberlake

By Stephen Kieran and James Timberlake

Cellophane House makes no claim to permanence. As a structure it is, first and foremost, a matrix for holding materials together in such a way that they create an inhabitable enclosure. The key term here is 'holding', as opposed to 'fixing'. Materials that are held are allowed to retain their identity as discrete elements, and can be released at any time. When materials are fixed to one another, they become part of a composite structure, from which they can be freed only through the expenditure of great amounts of energy. The actual materials are, in a sense, irrelevant: it is the manner in which they are joined together that defines the essence of a structure.

In striving for permanence, conventional construction techniques (including the bulk of techniques used in contemporary prefabricated structures) fix materials to one another in such a way that they lose their individuality. A conventional floor might consist of hardwood flooring impregnated with polyurethane resin, nailed and glued to a subfloor of plywood (wood veneers bonded by adhesives), which is in turn nailed and glued to a set of wooden joists. The only way to retrieve the individual materials from such an assemblage is to essentially demolish them, a process as inelegant as it is wasteful, and from which the materials emerge as virtually useless.

Cellophane House, by contrast, is assembled out of discrete materials held in place using methods that are quickly and easily reversible. Here, the individual materials retain their integrity and can be extracted, exchanged, reused or recycled. This is a structure that can disappear as quickly as it appears.

above: The aluminium extrusions that form the frame of the house feature a T-shaped groove on each side, which allows the beams and columns to be fastened with 'dry joints' as opposed to being welded or cemented together. The T-groove provides the mechanism for a variety of friction connections. It serves as the negative receptor for positive T-shaped bars that join two vertical members and prevent them from moving out of alignment. Bolt-together steel connectors accommodate the various horizontal, vertical and diagonal joints. The T-groove is also used as a channel for operating the sliding doors within the house.

opposite: The lighting scheme takes into consideration the transparency of the building materials, while acknowledging the modularity of the structure. The translucent floors are uplighted with dimmable linear-line voltage LED modules which have a similar colour of light to household incandescent lighting and in this application require the same amount of power as is required for lighting a standard house. LEDs are generally up to twice as efficient as standard household lighting. Further, the uplighting is adjustable and allows the structural plastic ceilings to reflect diffuse light back down into the space while creating a luminous plane to walk on. To maintain transparency, the wires are channelled through the grooves in the aluminium structural members.

right: The house can be customised with a variety of pre-installed mechanical, electrical and plumbing systems. Quick connections between elements allow for minimal specialised work on site. The photovoltaic panels adhered to the NextGen SmartWrap™ gather energy from the sun and channel it to a battery array in the mechanical room on the ground floor. Water can be heated by a solar collector installed on the roof, flowed through a convective loop to a holding tank, and then distributed to localised point-of-use hot-water heaters in the bathrooms and kitchens.

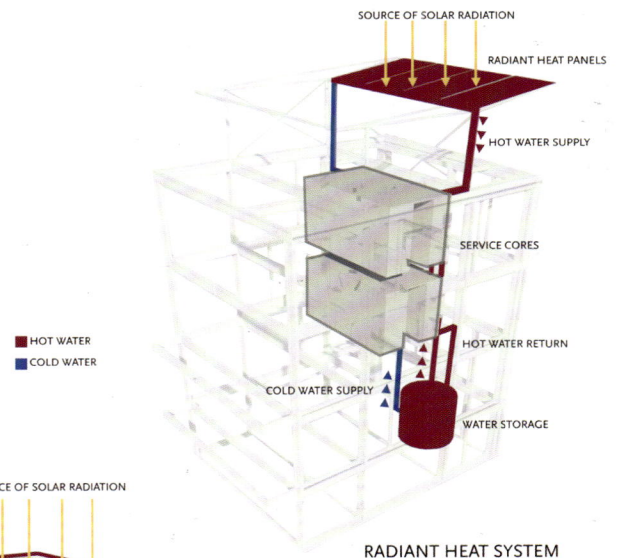

RADIANT HEAT SYSTEM

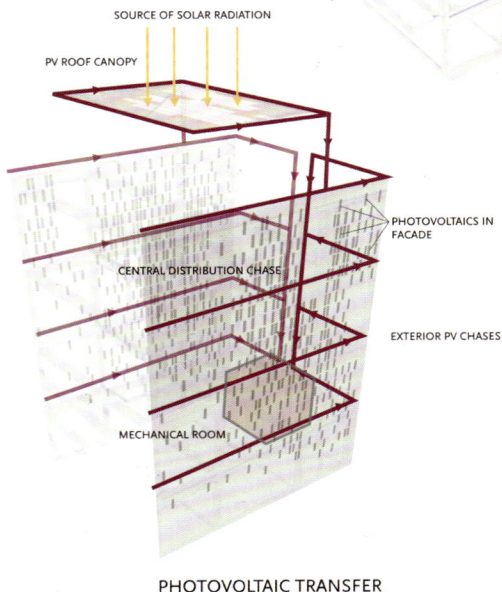

PHOTOVOLTAIC TRANSFER

Parallel-opening operable windows push out to allow airflow while remaining parallel to the building, providing ventilation without affecting the geometry of the facade. Schüco E2 glazing and NextGen SmartWrap™ are integrated with photovoltaic cells.

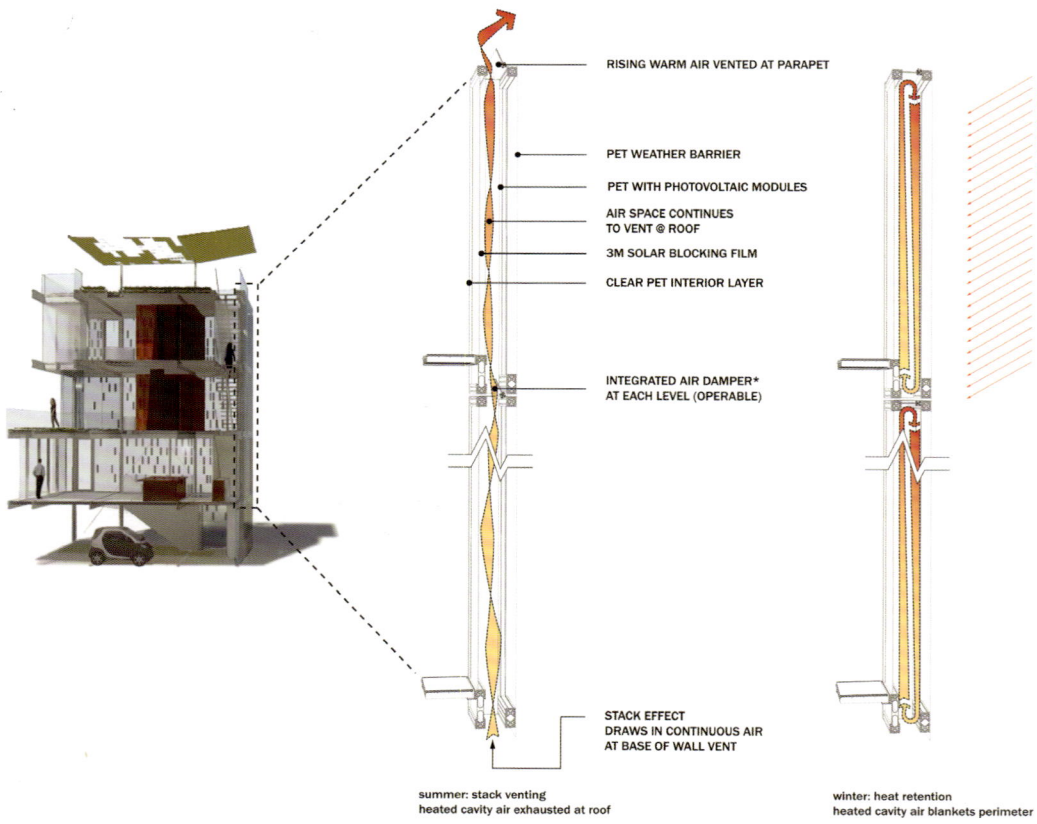

RISING WARM AIR VENTED AT PARAPET

PET WEATHER BARRIER

PET WITH PHOTOVOLTAIC MODULES

AIR SPACE CONTINUES TO VENT @ ROOF

3M SOLAR BLOCKING FILM

CLEAR PET INTERIOR LAYER

INTEGRATED AIR DAMPER* AT EACH LEVEL (OPERABLE)

STACK EFFECT DRAWS IN CONTINUOUS AIR AT BASE OF WALL VENT

SOLAR RADIATION

summer: stack venting
heated cavity air exhausted at roof

winter: heat retention
heated cavity air blankets perimeter

* dampers not deployed in version constructed for exhibit

The exterior walls of the dwelling are made from NextGen SmartWrap™ – consisting of an outer layer of transparent PET – the material used for soda bottles – and laminated with thin-film photovoltaic cells. The transparency allows sunlight to filter through the house, while solar power is harnessed through photovoltaic panels, enabling the house to function off-grid. An inner layer of 3M solar heat and UV blocking film lets daylight in while bouncing solar gain back out. A vented cavity between the two layers traps heat in the winter and vents it in the summer, reducing the amount of energy required to heat and cool the house. In addition, the south side of the dwelling features Schüco E2 glazing embedded with photovoltaic cells to give further energy independence.

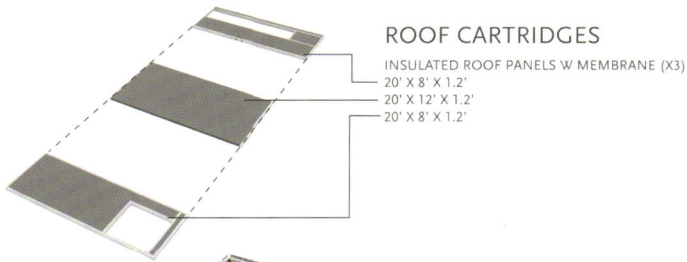

ROOF CARTRIDGES
INSULATED ROOF PANELS W MEMBRANE (X3)
20' X 8' X 1.2'
20' X 12' X 1.2'
20' X 8' X 1.2'

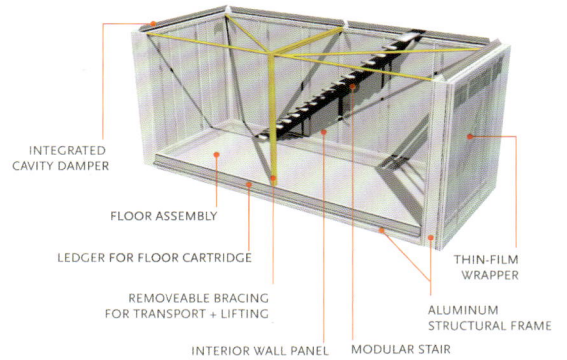

→ TYPICAL CIRCULATION BLOCK
(STACKED AT REAR)

INTEGRATED
CAVITY DAMPER

FLOOR ASSEMBLY

LEDGER FOR FLOOR CARTRIDGE

REMOVEABLE BRACING
FOR TRANSPORT + LIFTING

INTERIOR WALL PANEL   MODULAR STAIR

THIN-FILM
WRAPPER

ALUMINUM
STRUCTURAL FRAME

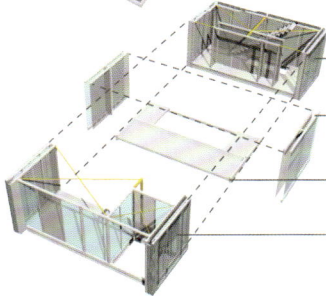

LEVEL 4
CIRCULATION/STORAGE BLOCK
20' X 8' X 10'

WALL CARTRIDGES (X2)
1.2' X 12' X 10'

FLOOR CARTRIDGE
20' X 12' X 1.2'

GLAZED BLOCK
20' X 8' X 10' H

SERVICES BLOCK
INTEGRATED BATHROOM STACK (2 LEVELS)
12.5' X 5' X 20'

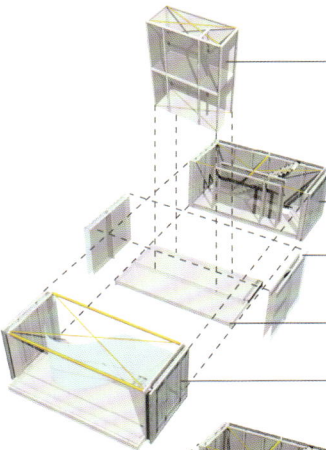

LEVEL 3
CIRCULATION/STORAGE BLOCK
20' X 8' X 10'

WALL CARTRIDGES (X2)
1.2' X 12' X 10'

FLOOR CARTRIDGE
20' X 12' X 1.2'

GLAZED BLOCK W/ THERMOFORMED CURVE
20' X 12' X 10'

= : TYPICAL FLOOR CARTRIDGE
(DROPPED IN BETWEEN PAIR OF BLOCKS)

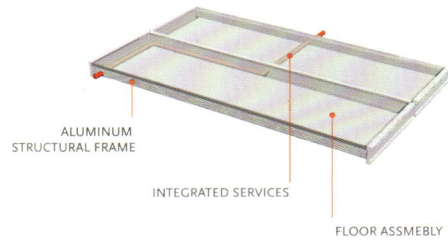

ALUMINUM
STRUCTURAL FRAME

INTEGRATED SERVICES

FLOOR ASSMEBLY

LEVEL 2
CIRCULATION BLOCK W/STAIR
20' X 8' X 10'

WALL CARTRIDGES (X2)
1.2' X 12' X 10'

FLOOR CARTRIDGE
20' X 12' X 1.2'

GLAZED BLOCK
20' X 12' X 10'

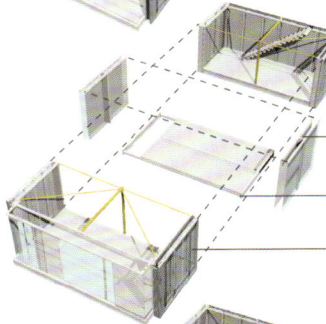

→ TYPICAL GLAZED FACADE BLOCK
(STACKED AT FRONT)

LEVEL 1
SERVICE/STORAGE BLOCK
20' X 8' X 9'

ENTRY STAIR BLOCK
4' X 12' X 9'

GROUND LEVEL FRAME

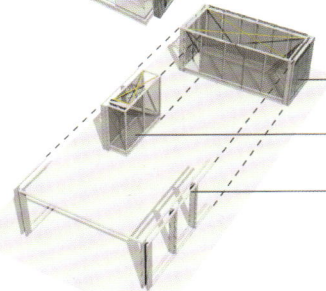

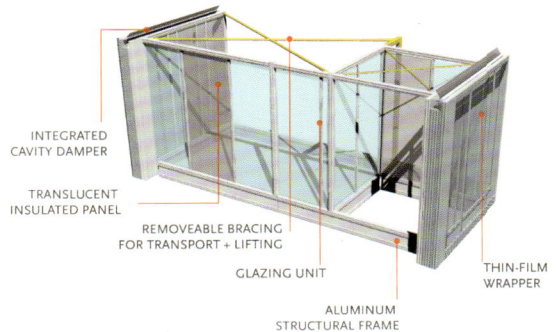

INTEGRATED
CAVITY DAMPER

TRANSLUCENT
INSULATED PANEL

REMOVEABLE BRACING
FOR TRANSPORT + LIFTING

GLAZING UNIT

ALUMINUM
STRUCTURAL FRAME

THIN-FILM
WRAPPER

Cellophane House is separated into logical parts, referred to as 'chunks',
which are integrated assemblies containing the aluminium frame, floors,
walls, bathrooms, stairs and exterior envelope.

Yale Sculpture Building and Gallery

New Haven, Connecticut, 2008

KieranTimberlake

By Stephen Kieran and James Timberlake

The Yale Sculpture Building and Gallery extends westwards across the university's extraordinary arts district, forging entirely new urban relationships with the city at the edge of the campus. Situated on a former brownfield site, the new complex invites the city into and through the site while providing perimeter street frontage where previously none existed.

The LEED Platinum Rating of this project was the outcome of a fully integrated design process that carried the project from programming through occupancy in 22 months, and relied heavily on information modelling. A four-storey glass studio building sits in the core of the perimeter block, with a single-storey store-front gallery on the street frontage to the north. Adjacent to the studio building, on the west side, is a four-storey parking garage with retail space on the ground floor. The gallery and parking structure re-establish the perimeter block, while the studio building acts as a lantern, illuminating the interior core. Mid-block pathways traverse the site in both directions, drawing pedestrians into the complex at the campus edge. The east–west path through the site is planned as an outdoor sculpture garden that stretches all the way back to Louis Kahn's existing Yale University Art Gallery.

The 260-square-metre (2,800-square-foot) sculpture gallery provides appropriately scaled street frontage adjacent to the historic houses along Edgewood Street. The gallery is clad in reclaimed wood to mimic the appearance of the existing houses, and the glass walls on its front fold away to become an open porch. An interior underground ramp connects the gallery back to the entry lobby of the studio building, and a sculptural steel stair zigzags from the basement to the third floor. Landscaped terraces on the first and second floors provide views to the gallery's green roof and the city beyond.

The 4,738-square-metre (51,000-square-foot) studio building contains three floors of individual and group studios above ground and basement floors of classrooms, machine shops and administrative spaces. Conceived as a loft building to accommodate a range of artistic activities, the structure has an exposed steel frame.

The large expanses of windows in each studio space provide 2 per cent daylight factor levels and views of the surrounding environment. Operable windows allow personal control over studio ventilation and daylight dimming ballasts are installed in all perimeter occupied spaces, subtly responding to interior light levels and providing full workspace lighting when natural lighting falls below 30 foot-candles.

SECOND FLOOR PLAN
SCALE 1:250

1. LOBBY / LOUNGE
2. PRESENTATION SPACE
3. STUDIO
4. EXTERIOR TERRACE

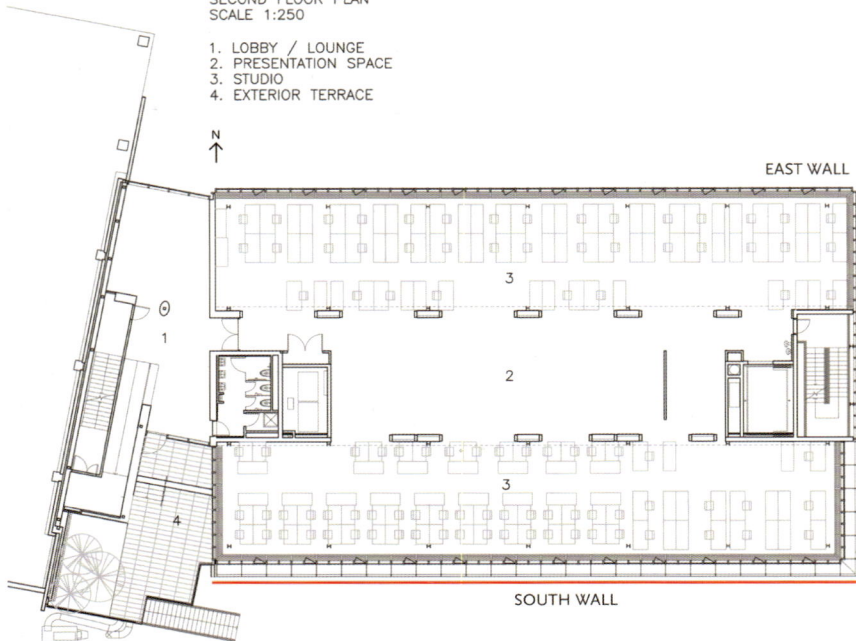

N
↑

EAST WALL

SOUTH WALL

To maintain a transparent envelope without compromising the building's high level of energy performance, a curtain wall made with transparent and translucent panels was combined with exterior sunshades to reduce solar heat gain. The windows are triple-glazed low-E glass, while the translucent spandrel panels, filled with the nanomaterial aerogel, are set within a glass frame. Two sets of spandrel panels on the south and east facades were monitored to determine the effect that venting the panels would have on cavity temperatures.

BRACKET

SUN
SHADE

LOW E
IGU

NANOGEL
KALWALL
(R=20)

FINTUBE

WINTER SOLAR NOON

SUMMER SOLAR NOON

72.5°

23.5°

Testing suggests that the overall R-value
(thermal resistance) of the spandrel assembly
is in excess of R 20 while maintaining 20 per
cent visible light transmittance.

The exterior sunshades on the south elevation that wrap around the
southeast corner of the building, and the three-bay organisation of
the plan, are apparent in the composition of the east elevation. ∆

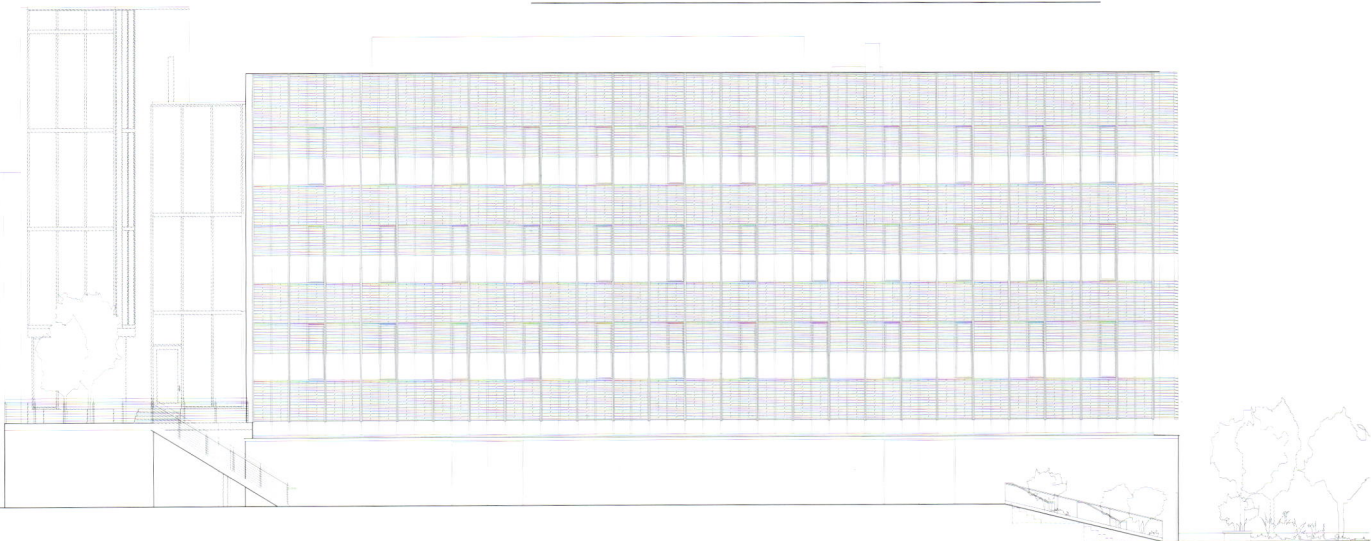

Building for the Third Century (B3C) Massachusetts General Hospital

Boston, Massachusetts, 2008-

NBBJ

By Craig Brimley and Jorge Gomez

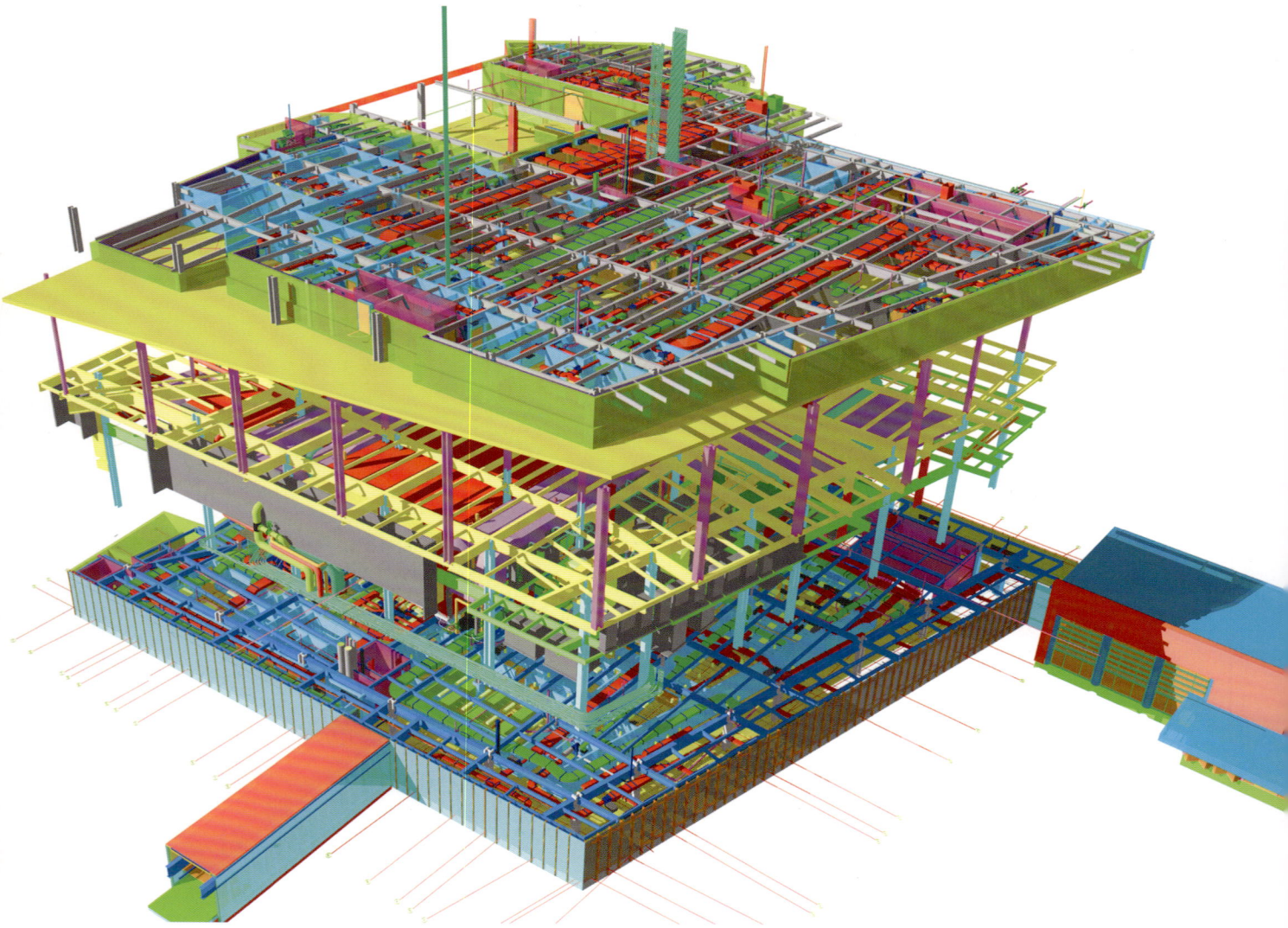

A redundant loop ventilation system was developed to allow complete interoperability of all the patient floors. The system enables the redistribution of air so that the various zones can be isolated to be adapted to future needs.

Due for completion in 2011 to mark the hospital's 200th anniversary, the Building for the Third Century (B3C) is situated in the midst of a dense urban medical campus in Boston's Beacon Hill district. Massachusetts General Hospital is one of the US's leading medical facilities, and the name of the new building reflects the continuation of its tradition of medical advancement into its next century.

As the cornerstone of the overall campus, the B3C building will be connected to five existing buildings from different eras. The design also needed to address future uses of the building. The design team thus employed building information modelling (BIM) technologies to deliver a project that provides the flexibility to incorporate new developments in medical equipment and that can adapt to the changing face of medicine.

The use of such a collaborative modelling process for the project was met with some reservations about the learning curve and the departure from traditional delivery methods. However, the promise of a quicker construction review, integrated delivery of design and a fully virtual understanding of the construction means and sequence of the building soon abated these concerns. The resultant information model included a multitude of documents such as 3-D diagrams, extractions and rendered perspectives, and was able to simulate performance with specific construction types, acoustic values and fire ratings. It also encouraged an ongoing discourse between the architects, consultants, client and the municipal review boards. As the model evolved, the design team was able to refine complex issues prior to construction. For example, collision detection algorithms helped mitigate major design changes on site as problems were identified and remedied early in the design process.

With systems designed to sustain the lifespan of the building, the coordination of the mechanical transition floor (centred between the patient and procedure floors) required special attention. Models of various building systems, such as structural and mechanical ducting, were imported into collision detection software where they were reviewed and major conflicts identified. An automated report was issued and reviewed in an on site interactive media room where the various subcontractors and fabricators involved in the project could walk through the model with the contractor and suggest immediate solutions to the conflicts identified.

With drawings tied to the model, a weekly workflow of design review, revision and model updates are immediately visible in the drawing set through an extraction process automated to run at the end of each work week. The following week restarts the process with updated 2-D documents ready to be reviewed and once again altered in 3-D model space.

New developments in the practice of medicine have brought about the opportunity to rethink the architectural organisation of hospitals, and their connection to the community. The B3C creates new links to five existing buildings, providing access to shared facilities, and extends its reach to the city with a link to the local metro station.

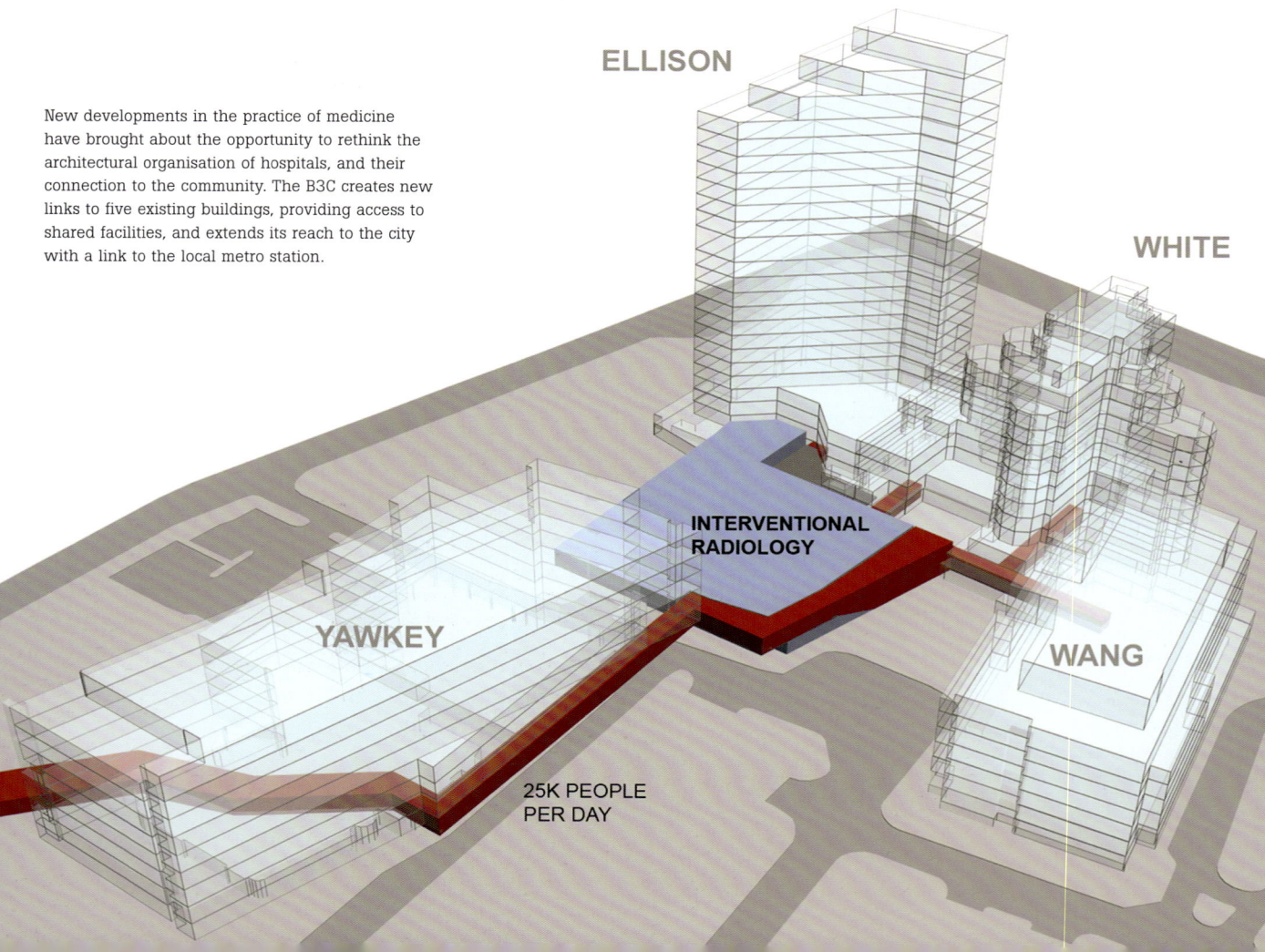

ELLISON

WHITE

INTERVENTIONAL
RADIOLOGY

YAWKEY

WANG

25K PEOPLE
PER DAY

The use of 3-D modelling was extended beyond the delivery of documents as a means to convey the design intent. The sketches, diagrams and visualisations produced illustrated the thought process of the designers and involved the client and municipal review board, which resulted in unanimous approval during the final project review, and allowed for the progression of unique ideas in medical architecture.

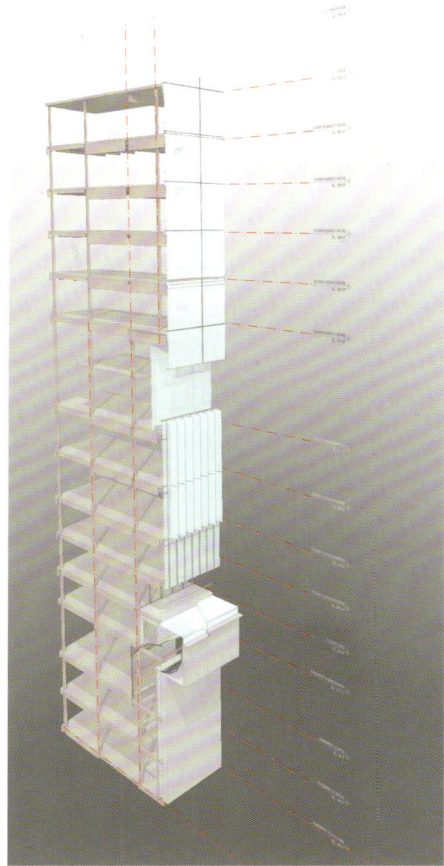

Through the coordination and revision of structural elements, sectional models were extracted from the overall model to visualise the impact of seemingly minor structural adjustments.

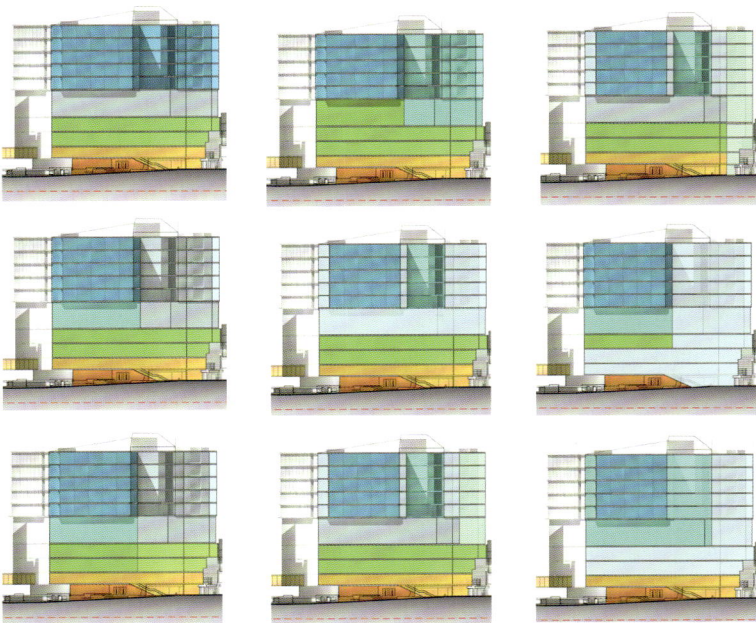

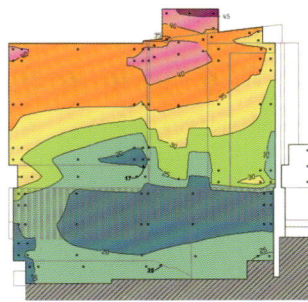

Relationships with adjacent buildings were studied in the development of the facade. As systems were designed, thermal dynamic models were tested to ensure the proper performance of exterior systems. ∆

Collaborative Intelligence =

KBAS, Pentagon Memorial, Arlington, Virginia, 2008
above: For several hours on sunny days, and from different angles, the custom non-directional satin finish of the stainless-steel Memorial Units catches dynamic reflections that move with the flowing water within the basins. As the units are oriented one way or another on the Age Lines, a pair of stainless-steel strips that run across the site parallel with the trajectory of American Airlines Flight 77 upon its impact with the Pentagon, so too does the directional flow of the water and reflections.

Shortly after 11 September 2001, a core group of family members related to the 184 individuals whose lives were lost at the Pentagon conceived of, and embarked upon, a mission to realise a memorial that would honour their loved ones. Eight months later, during the summer of 2002, an international design competition for the Pentagon Memorial was announced. Led by a team from the Army Corps of Engineers, the competition gathered together an esteemed group of jury members charged with judging the entries. Upon evaluating more than 1,100 entries, the jury awarded six finalists with a stipend to further refine their proposals. In March 2003, Kaseman Beckman Advanced Strategies (KBAS) was announced as the winner.

Adjacent to the point of impact of American Airlines Flight 77, the Pentagon Memorial is a place like no other.

Inviting personal interpretation on the part of the visitor, it provokes thought yet does not prescribe what to think or how to feel. Both individual and collective in nature, it intends to record the sheer magnitude of that tragic day by embedding layers of specificity that begin to tell the story of those whose lives were taken.

One hundred and eighty-four identical Memorial Units are organised by a timeline based on the ages of those lost at the Pentagon. They are uniquely placed along Age Lines that cross the entirety of the park – parallel to the trajectory of Flight 77 – with each line marking a birth year, ranging from 1930 to 1998. At the heart of the project, each Memorial Unit, highly articulate in its form and placement, is several things at once: it is a place to sit, a place for mementos, for contemplation, and the permanent epigraph of the victim. Inherent to the cantilevered form of the Memorial Unit, the orientation specifies whether an individual was aboard Flight 77 or in the Pentagon at the time of impact. Each name is engraved at the end of the cantilever,

Respect

Cultural projects today are often unprecedented in their complexity, scale and collaborative effort. **Keith Kaseman**, partner of Kaseman Beckman Advanced Strategies (KBAS), describes how building information modelling (BIM) enabled the realisation of the Pentagon Memorial in Arlington, Virginia. Twenty-eight contractors and consultants were brought on board in the design and delivery of the monument's 184 'memorial units', which collectively commemorate the individual lives lost at the Pentagon on 11 September 2001.

opposite top: The construction near to completion, as demonstrated by the landscaping and paving features installed in the Memorial Gateway, a 0.4-hectare (1-acre) zone that leads to the memorial proper. The cast-in-place concrete Age Wall is in the foreground, rising in height by 2.5 centimetres (1 inch) per birth year, and thus giving to drivers who pass by on the adjacent freeway the sense that some directional logic is in play. The view here is from the adjacent bicycle path, part of an extensive system across the whole of northern Virginia.

left: With plenty of time to spare before the official unveiling of the memorial, the Memorial Units were illuminated for the first time during a 4 am test. Using an electromagnetic induction lamp, the lights can operate for up to 75,000 hours before replacement bulbs are required. The custom light fixture is placed in a watertight dry box, the top of which is the weir over which the water flows, bringing leaves and other floating debris into a concealed filter basket. As it travels through the filter basket, the water is directed to hit the back of the light fixture box, removing the heat that has built up inside.

opposite right: Construction progress showing the Memorial Units distributed in the field. At this point, only the remaining 6.3-millimetre (¼ inch) thick finished gravel layer was yet to be installed on top of the multilayered 2.4-metre (8-foot) deep engineered fill. With the field of units established on a porous and true plane, all individual fountains needed to be balanced to ensure consistent water flow throughout the site. This not only ensures that the environmentally friendly water filtration and cleaning system works efficiently, but also serves to extract heat from the concealed heat sink of the light fixture at each individual unit.

hovering above a flowing pool of water that glows at night. Intensively custom-made, the Memorial Units were produced through complex procedures that interwove precision modelling, CNC production, analytical and finishing techniques.

On an operational level, KBAS has drawn many lessons from the intensive processes employed in the design of the Pentagon Memorial, particularly with respect to how digitally driven production facilitates fluid collaborative models throughout an expansive web of players and parameters. By cultivating and maintaining a critical rigour that permeated all substantive, non-standard production operations, including detail development and coordination of the logistics required to achieve the sublime intent of the memorial, numerous customised digital and physical protocols developed along each

nuanced branch of the collaborative network. Precision formed the core of this collaborative engine, in terms of geometric controls, tolerance envelopes and the communication of design goals. As such, massive efforts surrounding advanced modelling and fabrication procedures focused on information exchange, from digital to physical and back again, iteratively, through previously uncharted territory.

While thousands of people have contributed to the project at large, up to 200 have been involved specifically with the development of the Memorial Units. Manifested through extensive research and development, this element of the collaborative web consisted of 28 companies, fabrication shops, testing/research labs, organisations and consultants – all working to refine the interwoven techniques and processes necessary to produce the eight primary components that, when assembled, form each Memorial Unit.

By the time it was announced that KBAS had designed the winning scheme for the Pentagon Memorial, the PENtagon RENovation

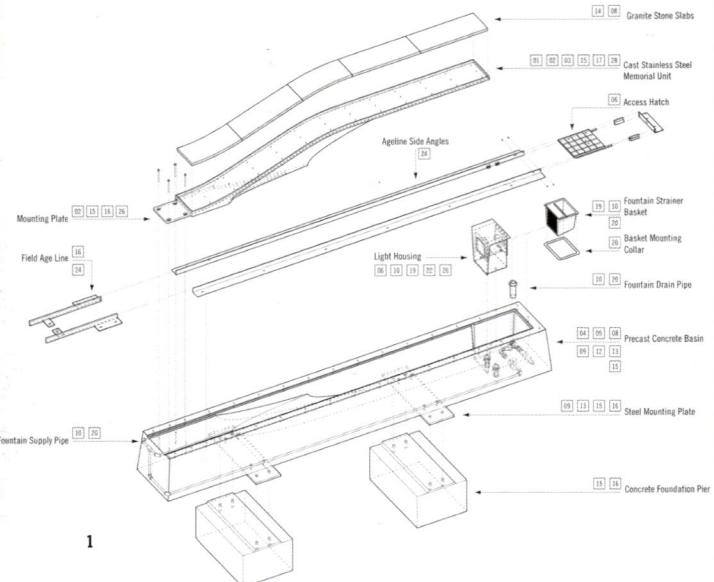

1

2

1. Twenty-eight companies, fabrication shops and organisations throughout the US were intensively involved with the development and production of the Memorial Units. Each component underwent dozens of iterations over the course of several years. High levels of precision were achieved through the use of CNC production techniques, and all physical connections were tested and refined through prototypes and mock-ups.

2. Casting mould components, made by compacting epoxy-bound sand into the production patterns at the Metaltek International foundry. A proprietary epoxy mix held the sand in place with the final consistency similar in hardness and texture to that of brick. The surfaces that were to come into contact with molten stainless steel were then given a smooth coating to lessen the number of surface impurities upon casting. Approximately 2,495 kilograms (5,500 pounds) of sand was used per cast, with each mould destroyed during extraction of the cast steel from within. All of the sand was recycled in the foundry for future use.

3. Stainless steel 316LN was cast into the production mould at a temperature of 1,607°C (2,925°F). The large vessels that carried the molten metal across the foundry to the moulds within which the casting took place were preheated to ensure that the proper temperature was maintained prior to pouring. If the temperature of the molten steel had dropped just a few degrees, it would have been necessary to begin the whole refined melting process again from scratch, which could have set operations back by up to 16 hours, as all casting took place on a particular work shift. The whole production sequence was refined through a prototype development process over the course of several years.

4. Heat treatment was required to realign the stainless-steel molecules to maximise corrosion resistance. During this process, the top half of the cast Memorial Unit was cleaned up (all gates, risers and any other remnants inherent to the casting process were cut or ground off) and then placed upside down in a large fixture made from a super-alloy. As the whole set-up was heated to more than 1,204°C (2,200°F), the stainless steel became pliable while the super-alloy, which remained hard as it requires much higher temperatures to reach a state of pliability, pushed the key features of the Memorial Unit into alignment slots and forms built into the fixture. Upon removal from the large furnace, precision cooling of the assembly was aided by cool mist blown by fans. Numerous alloys were considered for the Memorial Unit, however stainless steel 316LN was selected for its corrosion resistance properties and ability to refine the heat treatment process to a point where incredibly tight geometric tolerances could be achieved.

Program (PENREN), the public–private joint venture responsible for managing all construction projects on the Pentagon site, had already short-listed three design-build teams to compete for the contract to construct the memorial. KBAS worked closely with PENREN in developing the original Request for Proposal (RFP) from which the competing teams were to generate their bids and submit their proposals for developing the memorial. Given the magnitude and intensity of the endeavour, KBAS' input here centred on a call for extensive research and development, to be performed by a robust collaborative team of high-level experts across a broad design and fabrication spectrum. The contract was therefore awarded to Balfour Beatty/Lee Papa, a design-build joint venture between Balfour Beatty Construction (Fairfax, Virginia) and Lee + Papa and Associates (Washington DC), the landscape architecture/urban design firm and the executive architect for the memorial. From this point on, KBAS, Balfour Beatty and Lee + Papa were collectively positioned at the helm of what evolved into an incredibly diverse and extensive collaborative network through which the information models served as the primary conduit for developing all of the custom aspects of the memorial.

The R&D efforts began with the search for a foundry with the capacity to produce such a large cast with incredibly tight tolerance envelopes. Academic and industrial advisers with metallurgical expertise were appointed, and consulting engineers Buro Happold, with whom KBAS collaborated for the practice's original submission, continued to provide analysis and input throughout the development process.

Of only a handful of foundries in the US capable of producing the Memorial Unit cast, each differed significantly in terms of the alloys used, pattern types, casting techniques and tolerance management. Needless to say, the amount and quality of the knowledge mined through this phase was extraordinary, and all insight gained was incrementally fed back into KBAS' original information model of the Memorial Unit.

3   4

While the information model had to constantly be held within the parameters set forth by the design intent, KBAS was solely responsible for producing, refining, adjusting and maintaining all the computer models of the Memorial Unit. Wall thickness, draft angles, mass distribution and tolerances, for example, were contingent on the specific, intricate relationships at work throughout the various casting processes. Interestingly, this revealed certain technical principles regarding the behaviour of steel in the casting process, and in many cases this resulted in the production of an entirely new information model, rather than minor changes to an existing model.

As the formal and structural intent of the Memorial Units was already established in KBAS' original design, refinements through the R&D phase were an attempt to gain as much material efficiency as possible. The consistent influx of information and insight led to the systematic slimming down of the Memorial Unit from an original model that would have weighed 2,268 kilograms (5,000 pounds) to a refined version that weighed approximately 499 kilograms (1,100 pounds). Buro Happold performed finite element analyses on the refined model against load criteria developed by the KBAS team, which verified the structural performance at this stage of refinement. The final information model developed during this phase became the bid document for the RFP that was released to the foundries competing for the construction contract.

The Metaltek International foundry (Pevely, Missouri) was selected and brought under contract in May 2004. Immediately, the refined master model was transformed into a parametrically controlled solid model for pattern production, allowing Metaltek's engineers and Advanced Patternworks (Collinsville, Illinois) to modify the mould as necessary to facilitate mould production and an initial casting of the first prototype. Logistically complex by nature, the casting process was refined and adjusted through countless meetings and conference calls, and the first prototype was successfully cast in March 2005. However, upon testing out the procedures developed to take the cast Memorial Unit through the heat treatment process (to realign the molecular structure of the steel itself so that it has the capacity to perform as a corrosion-resistant alloy), it became clear that major procedural adjustments were required to prevent the unit from twisting and bending out of tolerance during this stage.

After months of intense brainstorming with all involved in developing the initial production strategies, and with new input from additional consultants, an intricate fabrication strategy emerged. Essentially, the top half of the units is cast stainless steel, while the bottom half (the fountain basin) is precast, self-consolidating concrete. This enabled incredibly tight tolerances to be achieved, even with the complex formal geometries, and ensured a level of adjustability through the installation process. Further, it launched the project into a chapter marked by amplified collaborative development and heightened coordination as the team expanded in both size and realms of expertise. As the new path forward was clarified, a whole host of additional digital file formats, material concerns and protocols, CNC production tools and analytical techniques emerged as integral facets of the complex, composite strategy that ultimately played itself out in realising the 184 Memorial Units now fully installed on the site. In the end, information from the final master model was translated from Rhino into formats suitable for use in Pro-E, Unigraphics, MicroStation, SolidWorks, AutoCAD, Adina, MasterCam and other customised applications required for specialised production. In numerous fabrication shops scattered across the country, the impressive array of large, specialised CNC tools used for production included five-axis mills, water jets, laser cutters, precision surface grinders, laser scanners and automated metal brakes.

Communication within the extended, collaborative network was primarily facilitated through the live navigation of the models, and within

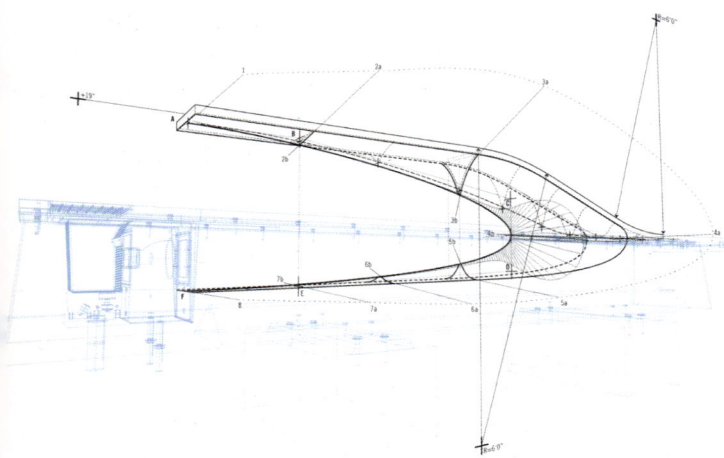

opposite: The field of Memorial Units creates an atmosphere fundamentally different at night than during the day. Multiple scales of engagement work simultaneously, from that of the passerby on the freeway or cycle path to that of the pedestrian visitor. In addition, as the Pentagon is adjacent to the flight path of Reagan National Airport, air passengers can view the illuminated memorial from above. The site is open to the public 24 hours a day, seven days a week; however, the most sublime experience to be had is perhaps in the middle of the night, when visitor traffic is minimal.

right: Refined, final production patterns awaiting mould construction for the next pour. Diligently refined, each pattern holds the geometry for two Memorial Units to be cast at the same time. The huge pattern components were fabricated by Advanced Patternworks, who used both incredibly large five-axis mills and standard hand tools to create patterns with a level of precision and durability sufficient for more than 92 casts. As the intricate casting process is inextricably tied to the geometries built into these original patterns, they will be stored in an undisclosed airconditioned facility owned by the Department of Defense in case a Memorial Unit needs to be replaced in the distant future.

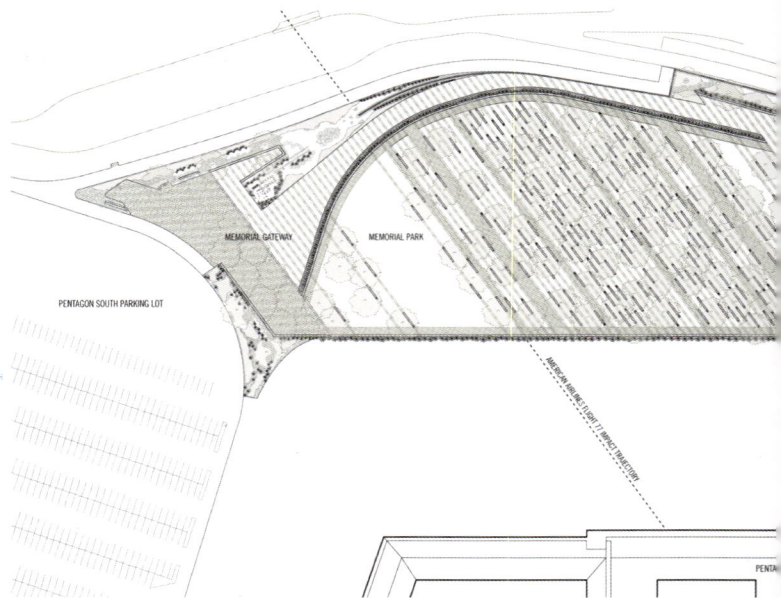

The geometric parameters of the final master model required dozens of iterations over the course of several years to fully refine. While the master information model was maintained through Rhino, digital versions of all components were translated into fabricator-specific software, including Pro-E, SolidWorks, Unigraphics and MasterCam, for analysis, prototyping and production. All were refined and produced via digital means, then fitted together with tight tolerances on site.

Adjacent to the point of impact of American Airlines Flight 77 on 11 September 2001, the Pentagon Memorial is an approximately 1.2-hectare (3-acre) park that invites personal interpretation from visitors. The individual Memorial Units are distributed across the site based on the ages of those who lost their lives that fateful day, creating a collective field that prompts contemplative thought.

this arena the expanded team of consultants were able to provide rapid input for further, specialised refinement. Through a prolonged flurry of information exchange and coordination, the insight gained was constantly translated into iterations of the model from which critical data was then extracted and sent back to the team for further scrutiny and feedback. The back-and-forth process of consultation and iterative refinement continued for almost three years, to the point where the mass collaboration reached a fluidly dynamic pace and anisotropy.

While the memorial contains a small collection of highly specialised fabricated components and details that are distributed across the larger field along the organisational Age Lines, all aspects of the Memorial Units and Age Lines were modelled at a high level of resolution. Throughout this whole development process, KBAS remained solely responsible for maintaining the master model of all components and orchestrating the information flows required for CNC fabrication. Numerous constituent components were translated from the master model into file formats applicable to the associated engineers' analysis and the fabricators' production means, often with the exertion of much more effort and manpower than anticipated. Upon successful translation, verified against the master model, each fabricator held

responsibility for, and ownership of, their own model, communicating refinements or suggested changes to the larger team via screenshots, drawings or interactive models. Once a refinement was collectively approved, pertinent data was fed back into the master model, extracted as required and distributed again throughout the collaborative network.

A key observation from this is that the more advanced the goal, the easier it is to close the gap between the productive promise of composite collaboration and deeply set construction norms. Given the magnitude of the challenge presented by the Pentagon project, and driven by a collective sense of urgency, information models facilitated the accumulation of precise knowledge across the collaborative network, surpassing the protocols inherent to contractual structures that typically bind a project team together. In this light, the Pentagon Memorial could serve as a model through which high-level collaboration, facilitated by nuanced information exchange and production protocols, is successfully demonstrated as uniquely productive. However, in the end such intensive activity was required simply to build a contemplative place where personal interpretation and reflection will persist through time. Officially opened on 11 September 2008, the Pentagon Memorial is open to visitors 24 hours every day, and the tracings of the collaborative flurry are quietly palpable in its capacity to elegantly radiate respect. ◬

Text © 2009 John Wiley & Sons Ltd. Images: pp 70 © Eddie Hidalgo; pp 71-4 © Keith Kaseman; p 75 © Melissa Kaseman

Her secret is patience and *She Changes*

Phoenix, Arizona (2009) and Praça Cidade Salvador, Porto, Portugal (2005)

Janet Echelman

By Janet Echelman

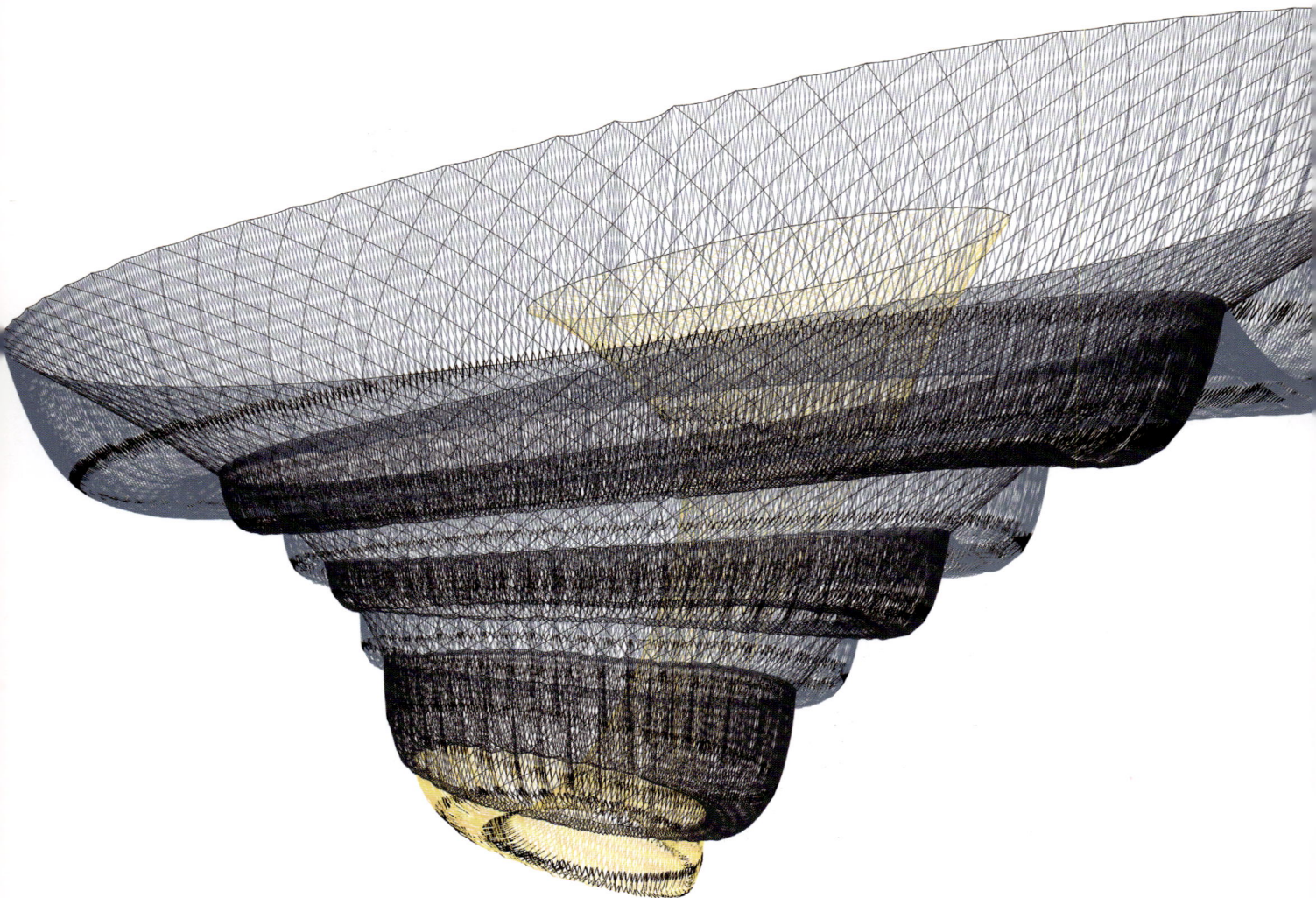

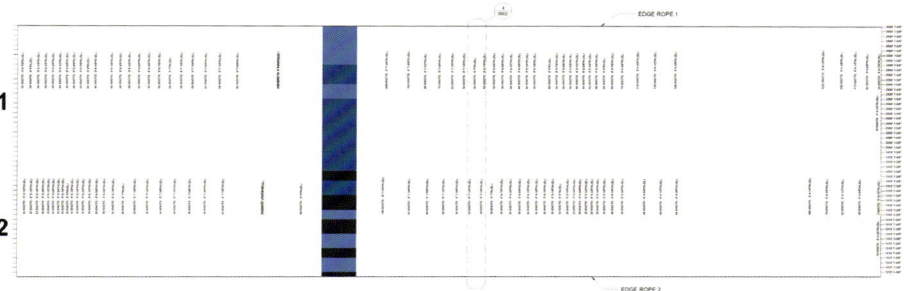

3 SCULPTURAL NET 1 FABRICATION LAYOUT
    SCALE: 1/4" = 1'-0"

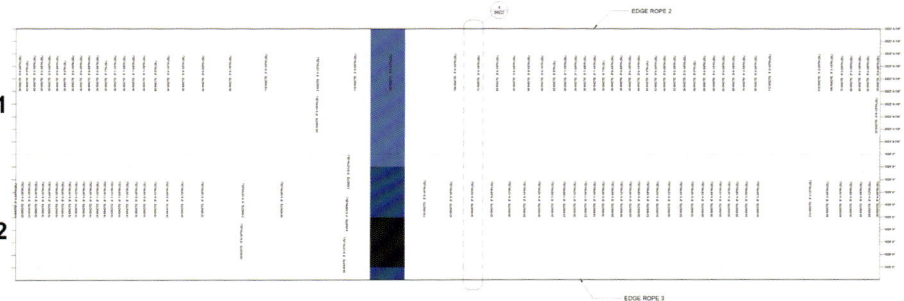

4 SCULPTURAL NET 2 FABRICATION LAYOUT
    SCALE: 1/4" = 1'-0"

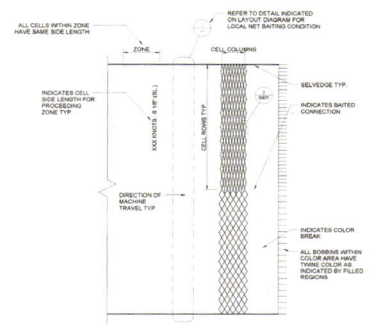

1 TYPICAL MACHINE KNOTTED NET LAYOUT DIAGRAM1
    SCALE: NTS

2 CELL DETAIL
    SCALE: 8" = 1'-0"

    COLOR LEGEND
    SCALE: NTS

Her secret is patience

above: Knotting details from traditional lace-making were employed to achieve scalar jumps from nets of one cell size to those with a smaller cell size. The baiting twine, shown in detail drawings 1, 2 and 4 overleaf is knotted by hand after machine fabrication of the adjacent net sections. The variation in net sizes along the vertical axis of the net panel allowed the artist to vary the opacity of the finished sculpture.

opposite: The overall form of the sculpture comprises seven sculptural net sections and an internal structural net. Gravity and wind loads are transferred through the net sections to the internal structure and then to a supporting steel armature. Virtual form-finding was performed on reduced-resolution models to verify the hanging shapes under dead load and to calculate reactions at the supporting structure. This 'dynamic relaxation' locates the minimal-stress shape for a given configuration of net, bar and node elements.

The intention has been to create sculptures in cities that interact with people in the course of their daily lives. Monumental, netted sculptural environments are animated by an ever-changing 'wind choreography', making the intricate patterns formed by this natural flow of air suddenly visible to the human eye as they are projected as shadow drawings on the ground below. The last decade has been spent developing a method for creating public sculptures that uses a rigid steel-armature system combined with flexible volumetric forms made of knotted high-tech fibre. The inspiration for these flexible membrane sculptures came from the design of life forms, in which a skeleton is draped with a skin to create a three-dimensional form. This was influenced by the study of evolutionary biology at Harvard University, and research into the ways surface area was maximised in a group of one-cell-thick life forms from the Precambrian era. For each project, the site, its geography, its physical, cultural and political history are researched, paying close attention to local materials, and how people have developed ways of working with such materials over the centuries.

A collaboration with Buro Happold Consulting Engineers, *Her secret is patience* for the Phoenix Civic Space, due for completion this year, transforms two city blocks in the centre of downtown Phoenix into a new civic icon in the sky. During the day, the sculpture uses sunlight to project the patterns of the Arizona Desert wind on to the paths of pedestrians. At night it is illuminated, and gradually changes colour throughout the seasons.

2005 saw the completion (with Eduardo Souto de Moura and Speranza Architects) of a voluminous net sculpture on the coast of Portugal. *She Changes* is a hollow form, 100 metres (328 feet) in diameter, that is suspended above a three-lane highway roundabout and moves in the wind. The sculpture has become the official symbol of the city. When interviewed, local people give different interpretations of the work: from the fishing nets, ships and masts of the country's maritime history and the red-and-white striped smokestacks of the area's industrial past, to Portuguese lace, sea creatures and ripples in water.

Her secret is patience

above: The Phoenix Civic Space sculpture makes visible to the human eye the patterns of the Arizona Desert winds. During the day, sunlight projects patterned shadow drawings on to the paths of pedestrians. The large, three-dimensional, multilayered form was created by combining hand and machine knotting. At night, the sculpture is illuminated, and the colour of the light changes throughout the seasons with an inverse relationship to temperature. In winter, the colours become hot reds and fuchsia, and in summer cool greens and blues.

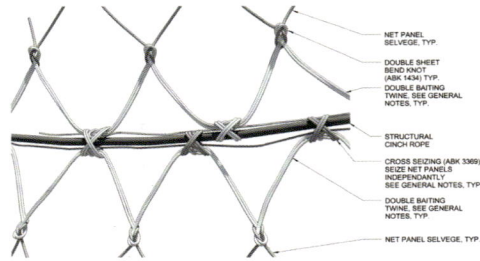

NET PANEL SELVEGE, TYP.
DOUBLE SHEET BEND KNOT (ABK 1434) TYP.
DOUBLE BAITING TWINE, SEE GENERAL NOTES, TYP.
STRUCTURAL CINCH ROPE
CROSS SEIZING (ABK 3369) SEIZE NET PANELS INDEPENDANTLY SEE GENERAL NOTES, TYP.
DOUBLE BAITING TWINE, SEE GENERAL NOTES, TYP.
NET PANEL SELVEGE, TYP.

① **STRUCTURAL NET CINCH ROPE CONNECTION DETAIL**
SCALE: NTS

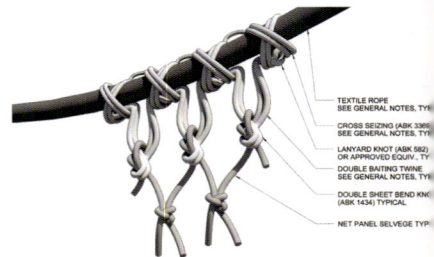

TEXTILE ROPE SEE GENERAL NOTES, TYP
CROSS SEIZING (ABK 3369) SEE GENERAL NOTES, TYP
LANYARD KNOT (ABK 582) OR APPROVED EQUIV., TY
DOUBLE BAITING TWINE SEE GENERAL NOTES, TYP
DOUBLE SHEET BEND KNO (ABK 1434) TYPICAL
NET PANEL SELVEGE TYP

② **TEXTILE ROPE NET CONNECTION DETAIL**
SCALE: NTS

right: Custom knot types were developed to avoid the problems of fatigue experienced in previous tests. A series of edge connection details were designed to ensure structural redundancy in the event of local failure of a net panel. The knotting techniques were also designed to facilitate ease of installation of the net panels. The knots were virtually tested with modelling software to understand how they would behave under stress. Edge ropes were digitally designed with UV-resistant jackets that prevented breakdown of the material due to the intense desert sun in Arizona.

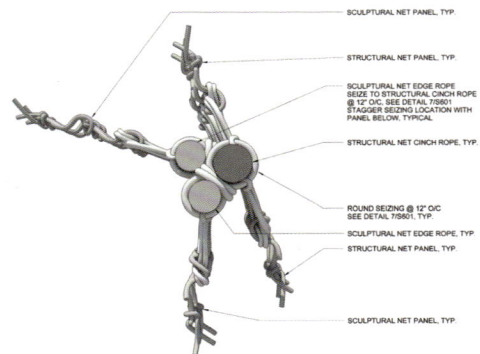

SCULPTURAL NET PANEL, TYP.
STRUCTURAL NET PANEL, TYP.
SCULPTURAL NET EDGE ROPE SEIZE TO STRUCTURAL CINCH ROPE @ 12" O/C, SEE DETAIL 7/S601 STAGGER SEIZING LOCATION WITH PANEL BELOW, TYPICAL
STRUCTURAL NET CINCH ROPE, TYP.
ROUND SEIZING @ 12" O/C SEE DETAIL 7/S601, TYP.
SCULPTURAL NET EDGE ROPE, TYP.
STRUCTURAL NET PANEL, TYP.
SCULPTURAL NET PANEL, TYP.

③ **CINCH ROPE CONNECTION DETAIL**
SCALE: NTS

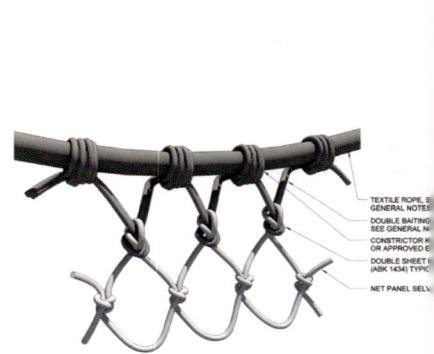

TEXTILE ROPE, S GENERAL NOTES
DOUBLE BAITING SEE GENERAL N
CONSTRICTOR K OR APPROVED E
DOUBLE SHEET (ABK 1434) TYPICAL
NET PANEL SELV

④ **TEXTILE ROPE NET CONNECTION ALTERNATE DETAIL**
SCALE: NTS

She Changes

Using colour and material methods to invoke memories of the site's historic past as a fishing and industrial centre, *She Changes* is a three-dimensional, multilayered net that floats over the area's waterfront plaza. This $1.6 million work, a collaboration with Eduardo Souto de Moura and Speranza Architects, and consulting engineers AFA Associates and Peter Heppel Associates, is credited as the first permanent, monumental public sculpture to use an entirely soft and flexible set of membranes that move fluidly in the wind. This casts cinematic shadows over the ground below. At night the sculpture is illuminated, becoming a beacon along the Portuguese coast.

The net is made up of 36 individual mesh sections in different densities, hand-joined along all sides to create the sculpture's multilayered form. The material used is GORE Tenara® architectural fibre, a 100 per cent UV-resistant, colourfast fabric made of PTFE (polytetrafluoroethylene), which is more commonly known as the non-stick cooking surface Teflon®. ∆

Information Modelling as a Paradigm Shift

Building information modelling (BIM) is not just a change in software or skills sets, it requires a paradigm shift. **Dennis Shelden**, Chief Technology Officer of Gehry Technologies, outlines the more 'fundamental, subtle and profound decisions' on the road to BIM. It is necessary to fully consider not only the impacts both 'upstream' and 'downstream' from the conventional design phase, but also the possible creative restrictions as there is a potential trade-off that comes with the emphasis on collaborative processes.

Wong & Ouyang with Gehry Technologies, One Island East, Hong Kong, 2008
BIM was used extensively for the construction planning and project controls for the One Island East project. These applications included detailed simulation of on-site construction activities. Particular focus was placed on simulating the operations of formwork lift sequencing, as table formwork for pouring cast in place slab and core elements were planned before construction took place.

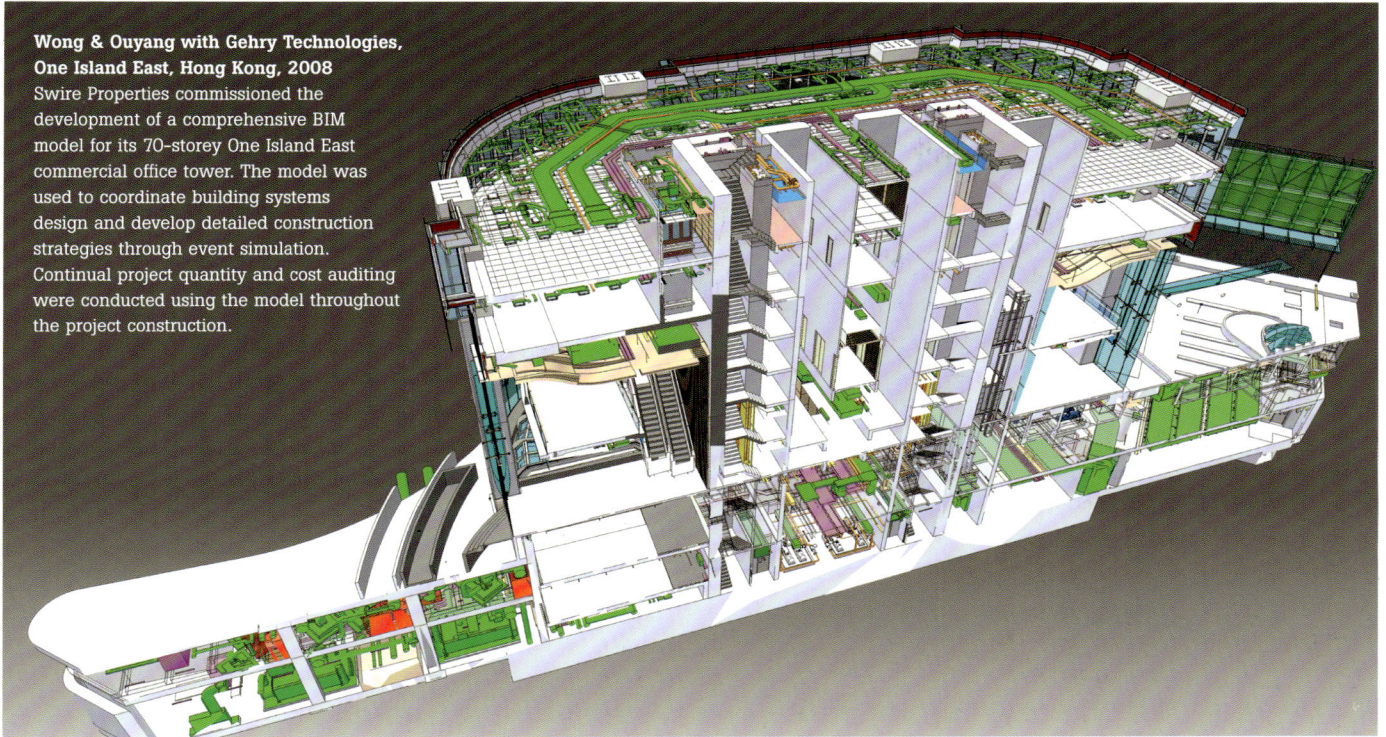

Wong & Ouyang with Gehry Technologies, One Island East, Hong Kong, 2008
Swire Properties commissioned the development of a comprehensive BIM model for its 70-storey One Island East commercial office tower. The model was used to coordinate building systems design and develop detailed construction strategies through event simulation. Continual project quantity and cost auditing were conducted using the model throughout the project construction.

Building information modelling (BIM) can be thought of as a paradigm shift from the traditional architect–contractor delivery process that is much more broad than simply selecting alternative software packages, or even 'doing the same thing but in 3-D'. As an evolution of building, information modelling has a wide impact on all practices, affecting what each party does, the authority and responsibility they take, and how they are compensated. Though the impacts of this paradigm shift are substantial, the evolution can be incremental as firms can choose how far they take this migration and at which points in time.

Design consultants are rightfully concerned with the cost/benefit and risk impacts of BIM. These concerns are partly the simple questions of what software products to invest in and what effect this decision will have on their practices. However, they are aware that there are more fundamental, subtle and profound decisions on the road to BIM, including:

- Which staff will be affected by a shift to information modelling and will existing staff be able to make the transition? This issue is perhaps far more critical as it has an impact on senior staff who will not necessarily be performing production work but will need to lead teams in the use of these new systems.
- Will the practice be taking on additional work or liability without additional compensation, and does its existing setup, including standard contracts and insurance policies, support the use of information

modelling packages? Are there risks in the adoption process of 'getting BIM wrong' on projects where it is applied?
- And, ultimately, how does BIM affect the ethos of the firm? Will there be quality or stylistic impacts, either positive or negative, and can design leadership drive the products of BIM to the outcomes that reflect the values of the firm? The concern may be that BIM will require (as opposed to enable) firms to take on additional aspects of professional project delivery beyond their core ambitions, or that the 'invisible hand' of a tool's particular capabilities will direct the qualities of their work towards particular formal decisions.

One of the potent areas for innovation is the implications of BIM both around and between traditional domains of practice, and towards issues both upstream and downstream from conventional design considerations. Upstream activities are those that would traditionally take place on projects prior to conventional architectural design activities. These activities include *pro forma* and building programme modelling, even the potential for optimisation cycles to be run between building financial modelling, programme development and conceptual design. Similarly, there are really interesting potential applications of parametric modelling for urban and master-planning.

Downstream capabilities include those for consultants' models to be applied to fabrication and construction processes, from detailed component engineering and site component placement to project control procedures. There are also ambitions – largely unexplored – for information models to have a role in facilities management and operations. Embedded systems do not have a real role in architectural thought, even though digital systems are already embedded all around

Diller, Scofidio + Renfro with Gehry Technologies, Lincoln Center – Alice Tully Hall, New York, 2009
For this concert hall renovation project, the fabrication constraints of a wood veneer panelling system were brought forward into the project's design development. Maximum surface curvature metrics were established, and surfaces redeveloped to adhere to these geometric constraints in order to rationalise the project geometry.

us in the spaces that we inhabit, as consumer products. A mobile phone with GPS is now a spatial technology, embedded in space and increasingly aware and responsive to spatial context. Mobile phones are also now integrated with environmental systems, allowing occupants to control room conditions. Architecture really does not yet have a clear response to these technologies. But this is likely to change, and in the same way that BIM has been a catalyst for rethinking boundaries between building professions it is likely to become a means for rethinking the borders between built spaces and occupying products.

In thinking about upstream and downstream changes, it should be kept in mind that BIM affects the conventional sequencing of project development, often moving downstream considerations forward and streamlining the revisiting of early phase decisions later in the project. It offers, for example, the opportunity to incorporate detailed fabrication logics and associated

pricing strategies early in the design development. Similarly, programmatic changes can be affected later on while reducing the cost of modifying documents that have already been developed.

Gehry Technologies has been involved with what might be called the collaborative aspects of BIM. These include the potential for tighter iteration between design and engineering, for bringing fabrication logic and construction planning into design, and giving clients greater visibility and control. BIM offers the capacity to support reprocessing, repurposing and reinterpreting of design information outside of the narrow project contexts of the tasks for which information was originally developed.

Parametric and Generative Capacities of Information Modelling
Parametric technologies allow detailed logic of system component organisations to be encoded into generative approaches, so that this level of project understanding can be applied as a part of formal generative techniques. Additionally, metrics including simple engineering analytics such as solar gain or tributary area can be incorporated into parametric objects, allowing objects to self-analyse

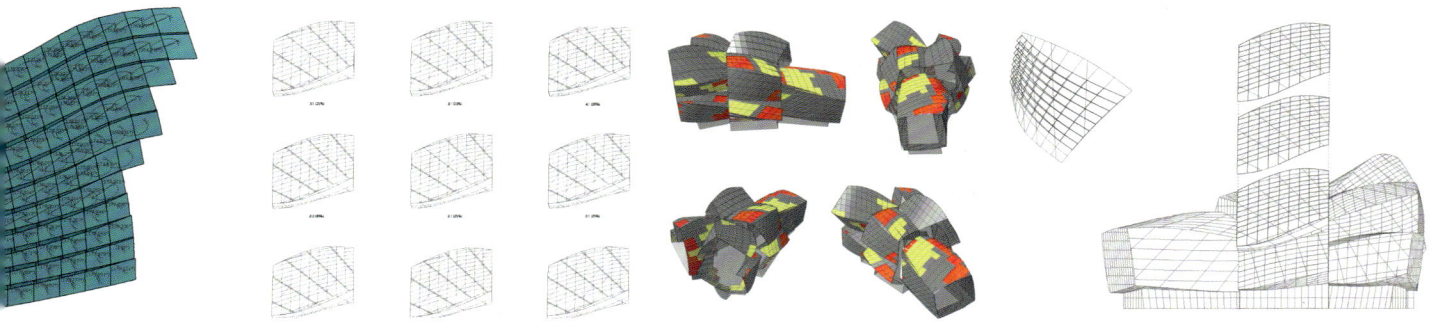

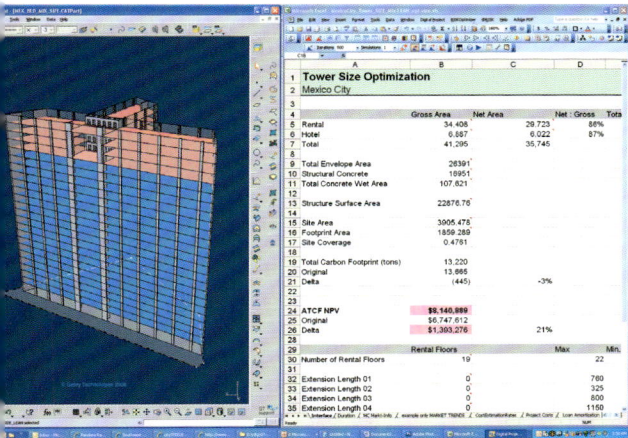

Gehry Partners, Gehry Building, NOVARTIS Campus, Basel, Switzerland, 2009
The Novartis Gehry Building is a part of a new campus master plan by the Italian architect Vittorio Lampugnani for the pharmaceutical company Novartis. Gehry Partners here used building modelling automation to develop rationalised enclosure panels. The algorithm drove the layout of panels on the surface, from which the strips of panels peeled away in order to satisfy the construction requirements of an assembly of flat quadrilateral panels. The algorithm fitted the panels to the surface and identified those that fell outside the construction requirements.

Gehry Technologies, BIM and Financial Optimisation, 2008
Building models can be integrated with project financial models and other planning-phase considerations. Here, a parametric building information model is derived from a spreadsheet-based financial *pro forma*. Quantity and occupancy metrics are sent back from the model to the *pro forma*. The integrated building information and finance model is iteratively modified to optimise the project's financial performance.

and self-solve. Despite the sophistication of these approaches, generative BIM can be tackled in a surprisingly incremental manner. The same algorithms that are used to instantiate general, formal shapes are used to generate systems with material or fabrication conformance. Some of the more trivial fabrication solutions, like triangulation, are no different than those generally used in computer graphics.

Authorial Creativity and Information Modelling

In a phenomenological sense, all design occurs as a feedback loop between what is existent and the act of forward projection. Design involving computational media should not be fundamentally different. Designers who take their inspiration 'from the screen' are adept at using these media as a context for exploration in a manner similar to – and often in conjunction with – traditional media.

However, all media carry certain unique affordances, and computation has of course unique characteristics that are different from traditional physical or worldly media. The most apparent such quality is that in order to be computationally operative, digital models must be built on constructs that are explicit, specific and consistent. This is not required of physical design media or operations on them – think of collage. There is something

of an inverse relationship between the efficacy or power of a computational approach and its flexibility. As a simple example, digital images are very flexible in capturing all sorts of design intentions, but the ability to computationally operate on this is very limited. Geometry and information model representations have higher-order operability, but are commensurably less flexible in what can be captured. The interesting questions this raises for design are: How much of the spectrum of potential design intentions can a given representation capture? Of this, how much is described in a manner that computers can 'understand', that is, be able to assist the designer in automating aspects of production while respecting intent? And finally, given the computational approach, what is required for a computer or human to translate the model to other sets of intentions?

BIM is broadly faced with this issue in that information models are expected to be operative across software and across disciplines. But in some ways this aspiration inherently requires us to restrict the set of potential intentions to the greatest common denominator of shared intentions, and there is an implicit trade-off between the support of the general and the unique. These are just some of the fundamental questions the BIM community is currently grappling with, although, of course, as technology development continues things can always get better. Δ

Strata Tower

Al Raha Beach, Abu Dhabi, United Arab Emirates, 2006–

Asymptote

By Hani Rashid and Lise Anne Couture

The tower resists being an overt, singular gesture reliant on a set meaning or association. Rather, its mathematical properties, not unlike those in the manifestation of the arabesque or abstract calligraphy, give the building its supreme elegance, prominence and significance. The twisting exoskeleton culminates with a roof canopy and helipad.

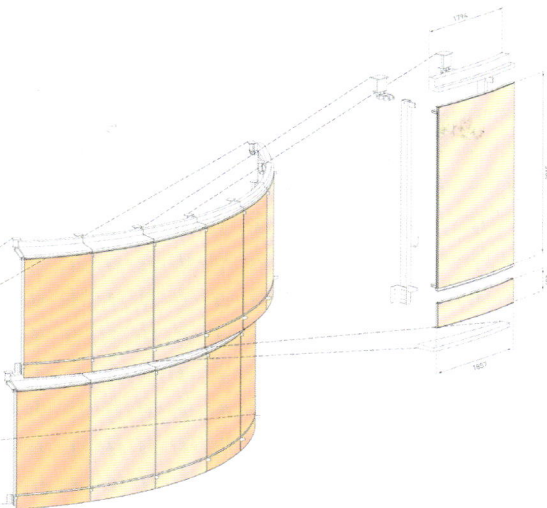

Currently under construction, and due for completion in 2011, at a height of 160 metres (525 feet) the Strata Tower, a 40-storey luxury residential building on Al Raha Beach, will be the tallest building in this exclusive waterfront development's Al Dana business precinct. Though the construction industry has traditionally resisted the use of building information modelling (BIM) technology, architects Hani Rashid and Lise Anne Couture of Asymptote here managed to convince the client to adopt these new tools in the development of the tower, which has resulted in better integration of the design and construction.

Parametric constraints were applied early in the project design so that different formal situations and possibilities could be developed, and internal volumes, floor areas and percentage of glazing to volume relationships maintained. The ultimate form of the building emerged from parametrically constraining the surfaces and volumes to respond to programmatic and environmental criteria. The design team is split into two groups: a 'front end' and an implementation group. The first was concerned with the geometric possibilities of the project and with what the BIM system allowed for in terms of creating new forms preloaded with intelligence and information. The latter group then embraced these as a virtual building en route to construction. This second group is also responsible for overseeing the client's needs as well as the construction process: streamlining, efficiencies, costs and so on.

The role of the architect has always been to innovate, experiment and produce polemical and pertinent solutions. BIM technologies are not intended to stifle such innovation or research into the sociocultural aspects of urban and architectural design. Instead, as in the case of the Strata Tower project, the aim is to enhance this through an integrated design solution that explores both new formal configurations and performance aspects of building in the Middle East.

Glazing types were optimised and component schedules generated through the information model. The tower glazing was parametrically modelled so that the surfaces could be rationalised into the maximum number of similar panels. These operations ensured the formal organisation of the building could be achieved while still adopting the efficiency of standardised parts. Schedules of dimensions from all of the glass and framing components could then be generated from the model. The diagram shows seven panel types, which vary in height and width. The model was also used to simulate fastening of the glazing panels to a metal frame on the building's exterior.

Interior view of the West Lounge. The Strata Tower's lobby features three floor-to-ceiling, highly reflective sculptural centrepieces that mirror the activity in the lobby as well as the movement of the surrounding seascape.

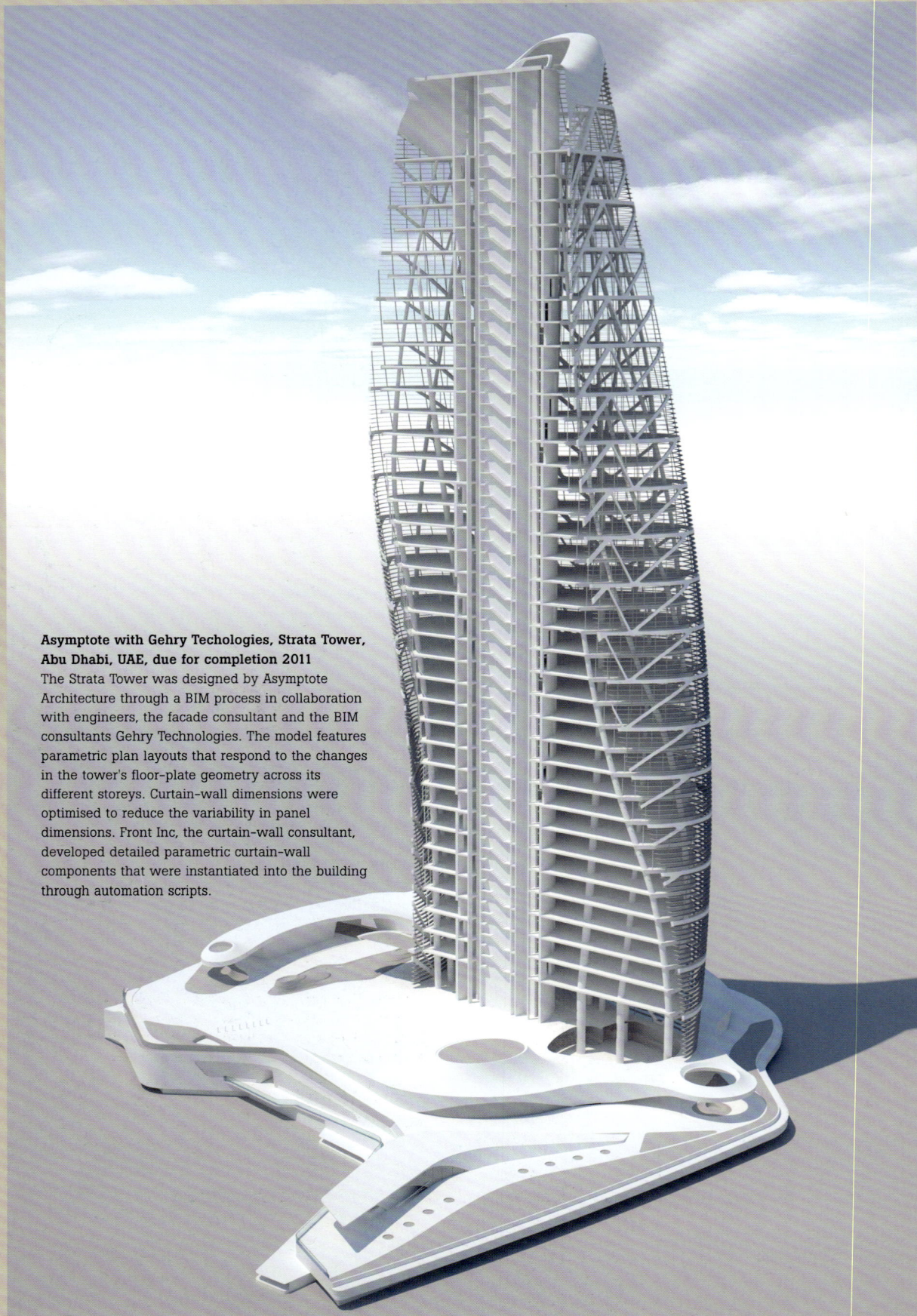

Asymptote with Gehry Techologies, Strata Tower, Abu Dhabi, UAE, due for completion 2011
The Strata Tower was designed by Asymptote Architecture through a BIM process in collaboration with engineers, the facade consultant and the BIM consultants Gehry Technologies. The model features parametric plan layouts that respond to the changes in the tower's floor-plate geometry across its different storeys. Curtain-wall dimensions were optimised to reduce the variability in panel dimensions. Front Inc, the curtain-wall consultant, developed detailed parametric curtain-wall components that were instantiated into the building through automation scripts.

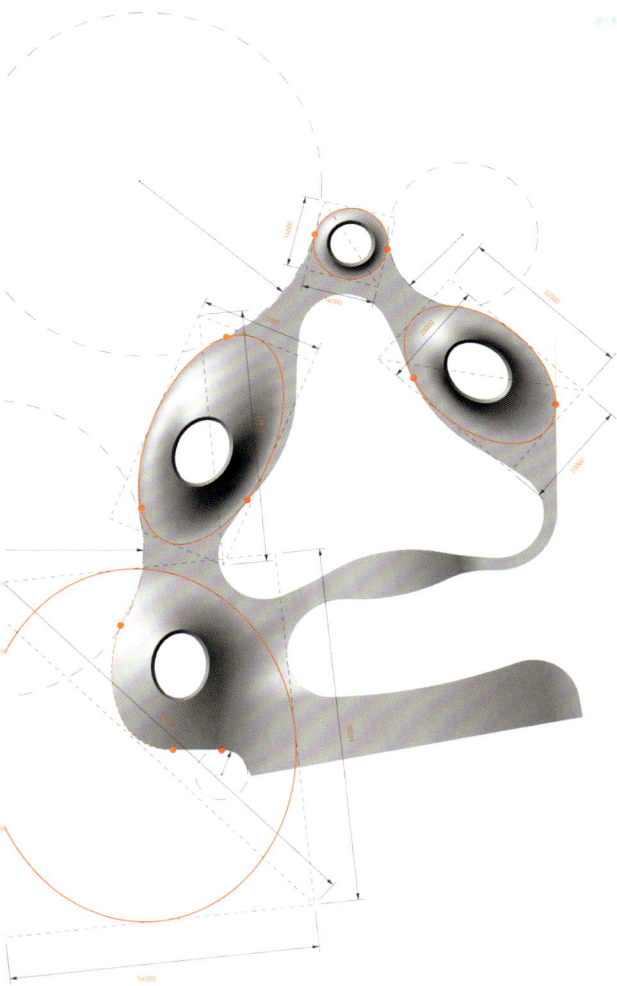

The roof canopy was parametrically modelled to allow iterative adjustments to the perimeter and multiple skylight oculi. Constraining the radial relationships between openings allowed design flexibility, but also fixed specific geometric relationships on the surface.

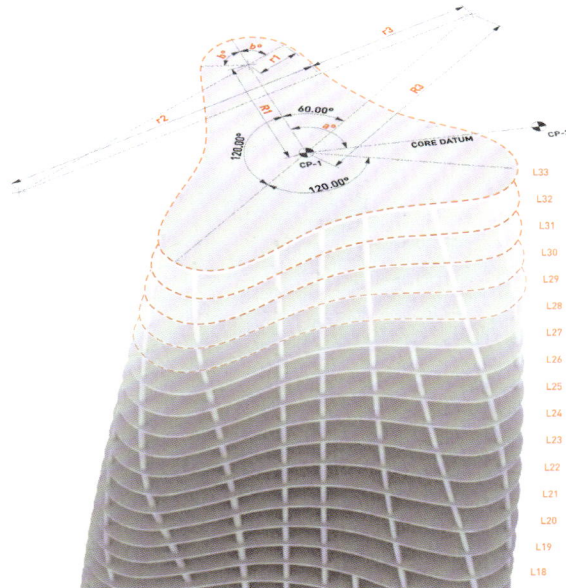

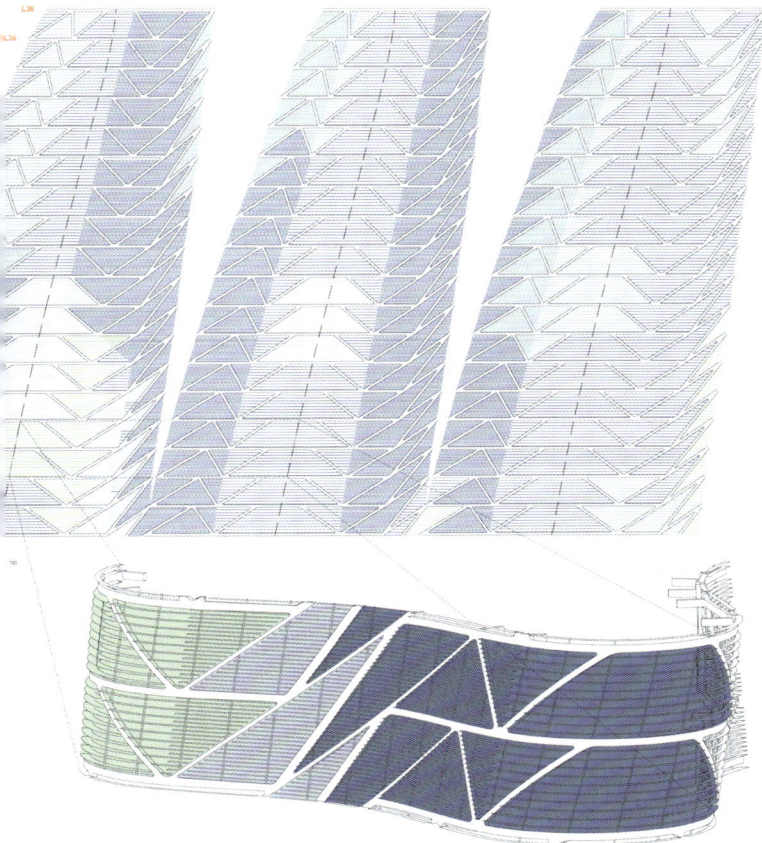

	a°	b°	R1	R3	r1	r2	r3
				SLAB GEOMETRY SETOUT			
L33	151.090	70.243	16988.442	34799.428	6056.151	71228.518	22957.234
L32	149.136	67.987	17435.135	35198.030	6168.467	41192.919	23137.684
L31	147.182	71.005	17869.936	35603.187	6284.342	78073.478	23351.839
L30	145.228	71.388	18293.012	36012.461	6369.373	82276.566	23570.819
L29	143.274	71.629	18704.508	36423.860	6506.901	84591.328	23778.547
L28	141.320	69.077	19104.547	36835.746	6592.560	46017.049	24001.223
L27	139.366	69.310	19493.360	37248.761	6698.627	47184.974	24223.229
L26	137.412	69.616	19870.682	37655.721	6790.112	48699.455	24462.985
L25	135.458	70.053	20234.959	38061.823	6859.272	51263.827	24729.649
L24	133.504	72.179	20892.144	38464.115	7082.446	58783.846	25040.844
L23	131.850	72.635	20936.304	38861.967	7078.772	89432.477	25170.012
L22	29.597	70.534	21269.500	39254.800	7150.098	54129.389	29606.950
L21	127.643	72.739	21591.785	39642.122	7291.395	89360.521	25582.123
L20	125.689	72.876	21903.056	40025.512	7380.078	91429.023	25806.669
L19	123.735	74.639	22203.808	40399.611	7556.296	103102.116	26137.389
L18	121.781	74.676	22493.630	40767.110	7669.682	102089.772	26362.840

The concrete structure and floor perimeter shapes of the tower were modelled with adjustable parameters for the height, taper and twist of the building envelope as well as variable floor-to-floor heights. The radial grid of sloping columns could be automatically updated following modifications to these parameters, updated manually in the virtual model, or numerically in a series of schedules (shown to the right of the diagram) that contained geometric information about each floor and column component. This allowed the tower form to evolve fluidly, as more information about specific programmes and building systems was attained.

Information modelling analysis and simulation was used to create the building's intelligent, environmentally responsive louvre system that is housed in a unique, cantilevered exoskeleton structure. An operation to unfold facade levels over floors 4 to 38 rationalised the size and distribution of the sun-shading louvres and integrated them within the building facade. The louvres were modelled with variable depth and distribution across the facade's curvilinear form according to environmental performance studies that focused mainly on Abu Dhabi's intense solar gain. ⚖

ALBERTI'S PARADIGM

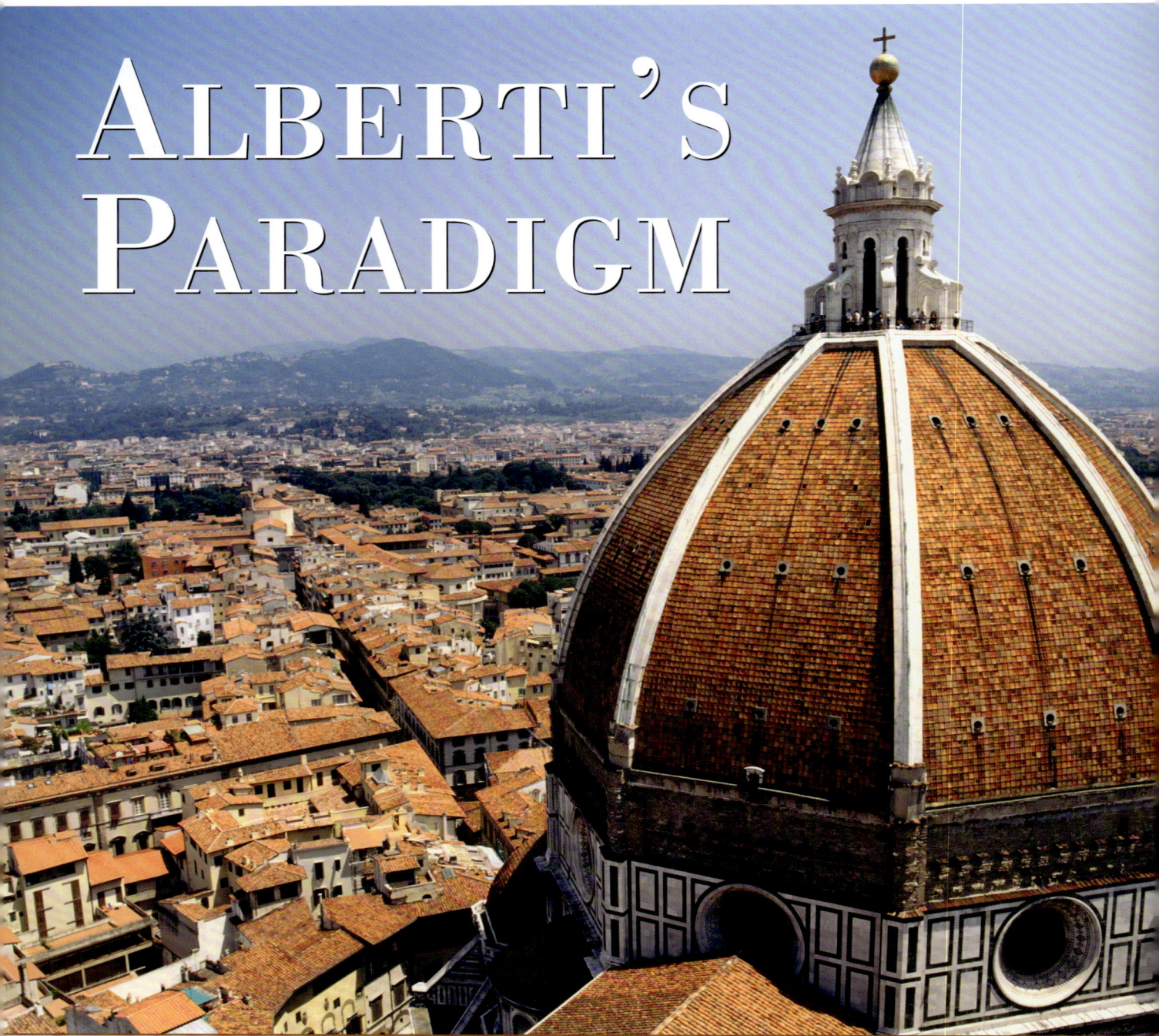

Could the advent of building information modelling (BIM) see the architect's status return to that of his medieval counterpart, the master builder? Guest-editor **Richard Garber** examines the work of Filippo Brunelleschi, who did the seemingly impossible in early 15th-century Florence by spanning the massive opening of the dome of Santa Maria del Fiore, the city's cathedral, through an adept use of physical models.

Filippo Brunelleschi, Vaulting of the Dome of Santa Maria del Fiore, Florence, Italy, _c_ 1420–36
The completed dome soars over the Florentine skyline and can be seen for miles. A series of external apertures in the exterior cavity bring light and air into the central cavity, which it is possible to walk through, while others were used to accept the armature of exterior scaffolding used by masons and bricklayers during construction. The total height of the dome and lantern is approximately 114 metres (375 feet). The dome's diameter is about 42 metres (138 feet). It is estimated that it weighs 37,000 metric tons (81,571,037 pounds), and the number of bricks used in its construction was more than four million.

As design processes more fully make the transition to the digital, there has been an increased interest in drawing parallels with the analogue counterparts to such processes, in some instances stretching back as far as 600 years. One of the comparisons many have made in the advent of building information modelling (BIM) is that these new technologies will somehow return architects to the status of the master builder of the pre-Renaissance – that the strict division, or gap, between architects who design buildings and the builders who construct them is being closed. It was the *capomaestro*, or master builder, who facilitated the smooth flow between design and construction in many of the notable projects of the late Middle Ages. Unlike today's practitioners who graduate from the many dedicated architectural schools, the master builders of this period were most likely exceptionally talented craftsmen or artisans, still trained in guilds, who crossed over into architecture from an allied art. Their knowledge of making not only allowed them to conceive the design of buildings, but also gave them the opportunity to specifically formulate the construction sequence and engineer building practices.

Among the several notable projects of the 15th century was the vaulting of the massive dome of Santa Maria del Fiore (*c* 1420–36) in Florence by Filippo Brunelleschi. Brunelleschi was a trained goldsmith[1] who won prominence and acclaim for his daring and untested plan to construct the dome without the wood scaffolding traditionally used for centring. According to most interpretations, as the project's *capomaestro* he designed and engineered many innovative tools and techniques in the process that would form the basis for many of the built successes of the Renaissance.

Brunelleschi and Early Information Models

The most significant way new ideas about building and construction technologies could be examined during this period was through the making of large physical *modelli* crafted by designers to convey their intentions. These were in many ways similar to today's information models. They were usually large enough to be entered and inspected by clients, and they conveyed spatial organisation as well as information about materials and construction techniques. In the largest of these, craftsmen from the various guilds who would oversee the actual building were employed to construct the portions of the scaled model they would ultimately be responsible for in the field.

Brunelleschi's model for the dome of the Santa Maria del Fiore was some 18 metres (60 feet) in length and was constructed by four bricklayers lent to the architect by the Florentine Building Commission.[2] A series of wooden (and possibly stone) tension chains were set at equal distances

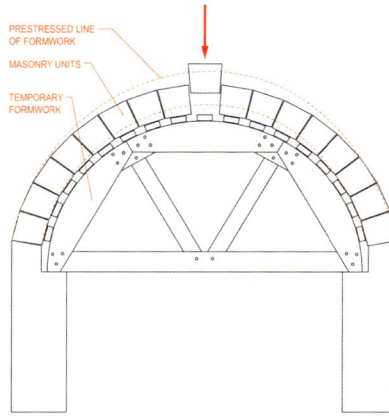

PRESTRESSED LINE OF FORMWORK
MASONRY UNITS
TEMPORARY FORMWORK

TYPICAL SCAFFOLDING FOR CENTERING A MASONRY ARCH

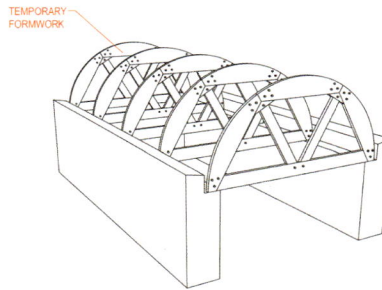

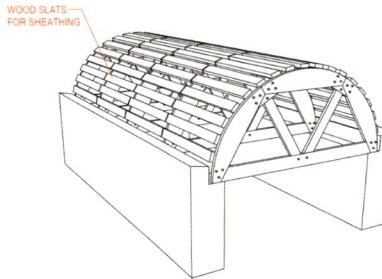

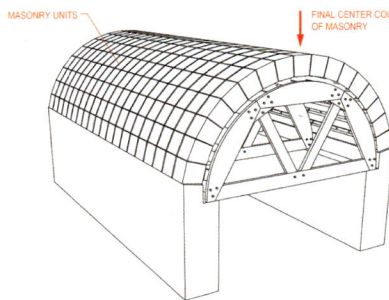

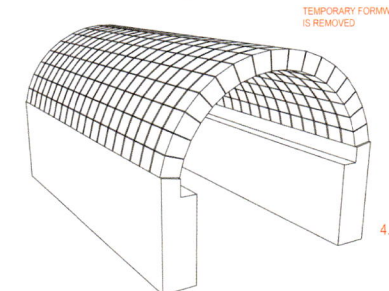

TEMPORARY FORMWORK

1.

WOOD SLATS FOR SHEATHING

2.

MASONRY UNITS

FINAL CENTER COURSE OF MASONRY

3.

TEMPORARY FORMWORK IS REMOVED

4.

TYPICAL SCAFFOLDING FOR CENTERING A MASONRY ARCH

The timber formwork required in vault or dome centring was typically stressed into place by the weight and fit of masonry units. Masonry units were applied from each side and culminated in a centre course that locked the units into their final location. As the units were installed, stresses on the temporary timber frame caused it to deform, thus it was critical to make the timber frame as rigid as possible to minimise such damage. Drawing based on illustrations by John Fitchen in *Building Construction Before Mechanization*, MIT Press (Cambridge, MA), 1989, pp 100–05.

Traditional Vaulting Drawings, 2008
The conventional method of vaulting an arch or dome prior to Brunelleschi's plan to vault the dome of the Florence cathedral was by first constructing and erecting a temporary support structure, usually of timber frame members, that was hoisted into place (1). Not only was the erecting and decentring, or removing, of these heavy members dangerous work, but deformation due to the settling of either the timber frame or masonry members also needed to be considered. Masonry was installed on top of a series of wooden slats that formed sheathing on top of the timber frames (2). Once the final centre course of masonry was installed, the centring could be removed (3). In many cases, the size of the dome or arch meant that the only practical way to remove the timber frame was to dismantle it piece by piece, high in the air – a daunting process in itself (4). Drawing based on illustrations by John Fitchen in *Building Construction Before Mechanization*, MIT Press (Cambridge, MA), 1989, pp 100–05.

within the interior cavity of the dome itself so as to invisibly support the structure, using neither buttresses, as seen in the Gothic cathedrals of the previous century, nor internal armature or scaffolding. The model thereby fostered a productive discussion and the opportunity for feedback from the various trades before the vaulting of the dome itself, and helped to ease the concerns of Brunelleschi's employers, the wardens of the Opera del Duomo,[3] as to whether such construction techniques would actually work. There is a clear correlation between the ability of such a working model to systemically test construction techniques and sequencing as opposed to simply representing a designer's intentions. In fact, some speculate that these models became more sophisticated as a more three-dimensional awareness emerged in the guilds. Similarly, virtual sequencing is made possible by information models today.

Christoph Frommel has noted that 'the architectural model must have evolved because of the same need for material and spatial quality and could have even been the response of the master builders to the illusionism of the painter-architects'.[4] The model also ensured a sort of architectural precision about the whole volume of the proposed building, if not its construction details. As the rise of the contemporary architect-designer separated the architect from the building site, models would have ensured a three-dimensional understanding of the scope of the work at a time when drawings did not figure prominently in the planning of buildings.

Trained in the guilds, Brunelleschi was a secretive and competitive man, and his architectural models of buildings and details were rarely finished so as to keep his design intentions covert. They lacked ornament, and only showed relationships between walls and other principal elements.[5] As many of these models initially served as competition entries for various parts of the dome, it is apparent why Brunelleschi wished to keep information from rivals such as Lorenzo Ghiberti who were also working on the project. In addition, it seems in keeping with his role as the master builder that Brunelleschi also kept information from workers on the site as a way of maintaining control of nearly every aspect of the construction. It was common for competition presentation models to eventually be used as guides for construction. As such, these models were routinely updated and added to as new technologies were developed and the construction of specific projects progressed.

Brunelleschi's use of such information models is consistent with their further employment in the 16th century, when a better understanding of construction technologies and higher degrees of craftsmanship meant that fewer changes to models were required once construction had begun. In contrast to the 15th-century

Filippo Brunelleschi, Vaulting of the Dome of Santa Maria del Fiore, Florence, Italy, *c* 1420–36
Model of the dome and apse of the Florence cathedral exhibited at the Palazzo Grassi in Venice in 1994. The exhibition, organised by Henry Millon and Vittorio Lampugnani and titled 'The Renaissance from Brunelleschi to Michelangelo: The Representation of Architecture', called attention to the importance of models in the design and construction of late-medieval and Renaissance works. Models were typically large enough to be entered by clients or tradesmen, and the master builder often employed the latter in their construction to ensure coordination of design intent with construction methods. Such models are generally thought to be more useful in the study and understanding of early Renaissance construction projects than drawings.

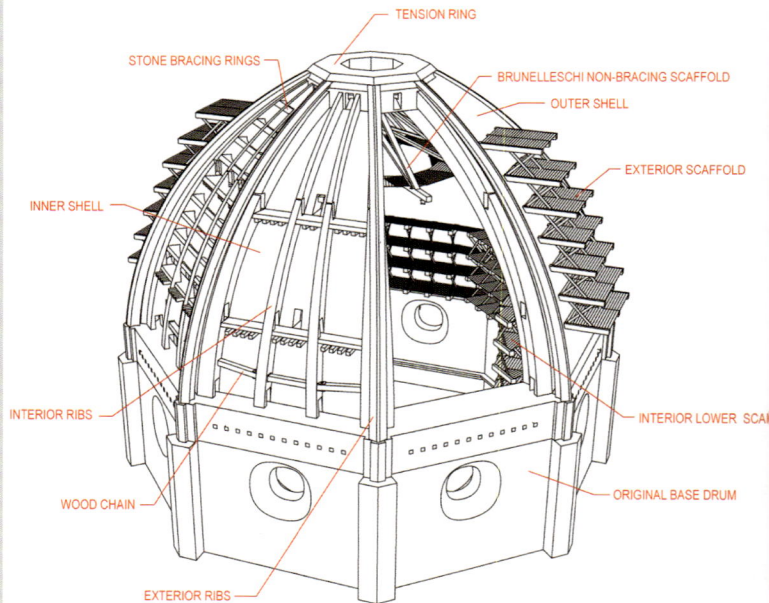

CONSTRUCTION OF THE DOME AT SANTA MARIA DEL FIORE, FLORENCE

The vaulting without centring of the Santa Maria del Fiore was a complex undertaking. Brunelleschi employed information models in order to better organise and convey his intentions to the masons and bricklayers who carried out this work. The numerous innovations in the construction of the dome included a two-shell system between which a series of wooden chain (and possibly stone) tension rings were installed within the cavity to resist the outward pressures of the form itself. This allowed construction of the dome without external buttressing or internal structural centring frames. In addition to the interior scaffolding, a series of exterior platforms were designed to fit into the exterior masonry of the dome. Drawing based on illustrations by Frank D Prager, in *Brunelleschi: Studies of His Technology and Inventions*, Dover Publications (Mineola, New York), 2004, p 35.

modelli, *modani*, or full-scale copy models (representations), of building profiles in wood were now used as a basis for the fabrication of details such as lintels or cornices.[6] These were like jigs, used for the replication of architectural features. It is interesting to note that *modani* began to emerge following the establishment of Gutenberg's printing press in Europe (1465), which helped to further codify architectural orders and disseminate knowledge about construction techniques through published treatises. They also seem to have been the precursors to 'mock-ups'.

Among Brunelleschi's most important inventions that revolutionised the construction of the dome was his oxen-driven hoisting device.[7] This machine was used to deliver materials several hundred feet in the air to the masons laying the dome's complex brickwork, and was significant because none of the previous pulley systems, driven by men, had ever reached such heights. At its retirement, it was estimated that the hoist had lifted 31.7 million kilograms (70 million pounds) of building materials to the masons and bricklayers.

Alberti and the Codification of Architecture

While Brunelleschi was working through the problem of vaulting the dome in the field, using scaled models and his advanced construction technologies, the theorising of such a scientific concept of art and building for the time was concurrently being undertaken by Leon Battista Alberti. In his treatise 'Della pittura', he wrote of Brunelleschi and his accomplishments: 'Who is so dull or jealous that he would not admire Filippo the architect, in the face of this gigantic structure, rising above the vaults of heaven, wide enough to receive in its shade all the people of Tuscany; *built without the aid of any truss work or mass of timber* – an achievement that certainly seemed impossible?'[8] It was also in this text that he would define modern principles of representation by geometrically defining the instrument of perspective and its use in architecture and design.

Alberti held various official posts with the papacy in Rome during his lifetime, while practising architecture and authoring his own *Ten Books on Architecture* (*De re Aedificatoria*) among other texts. *De re Aedificatoria* (1452) was a projective document in contrast to Vitruvius' *Ten Books* (*De architectura*), and sought to refine and develop the idea of architecture as a humanist art, the architect as an intellectual, and the practice of architecture as the loftiest of professions. The treatise is based on Alberti's survey of Classical orders and led to his calling for a more theoretical understanding of the use of these orders. Through his work with the papacy, Alberti was well travelled and had the opportunity to survey

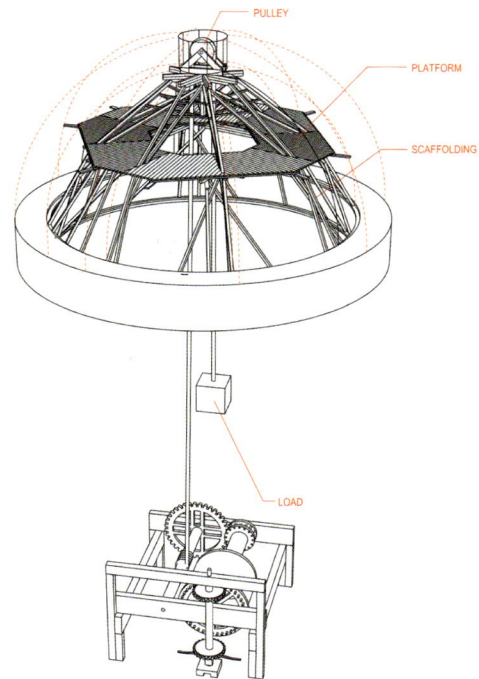

The relationship of the oxen-driven hoisting device invented by Brunelleschi and the scaffolding that was erected at the level of the dome's cupola was carefully coordinated. The scaffolding was not required to centre the dome – this was achieved through a series of chain rings and patterns of bricklaying, and as such was constructed of much lighter elements than the more conventional scaffolding of the period. At the top was a pulley that brought materials through the scaffolding to a series of levels where the tradesmen were working. At the base of the hoisting device were gears that allowed it to be moved forward or in reverse without changing the direction of travel of the oxen that were tied to it. Drawing based on illustrations by Frank D Prager, in *Brunelleschi: Studies of His Technology and Inventions*, Dover Publications (Mineola, New York), 2004, p 28.

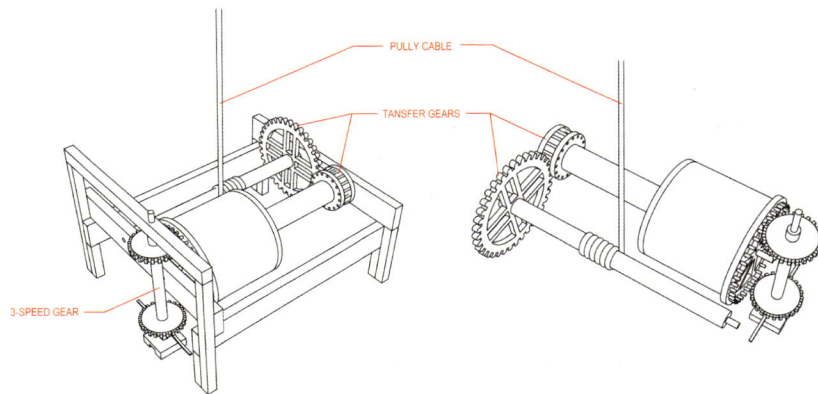

HOISTING DEVICE AND GEAR DETAIL

Powered by oxen, Brunelleschi's hoisting device involved a set of gears that moved differentially and in both forward or reverse by way of a reversing clutch and screw-controlled load positioner. It has been estimated that the hoist had lifted 31.7 million kilograms (70 million pounds) of building materials to the masons and bricklayers by its retirement. Frank D Prager suggests that the device was worked on by Antonio Manetti Ciaccheri, who was to become *capomaestro* after Brunelleschi. Drawing based on illustrations by Frank D Prager, in *Brunelleschi: Studies of His Technology and Inventions*, Dover Publications (Mineola, New York), 2004, p 71.

Classical works first-hand, which prior to the diffusion of the woodcut or the arrival of Gutenberg's printing press in Italy, was the only way to see such buildings. Interestingly, however, even contemporary translations contain very few illustrations, and as such are in keeping with the late-medieval practice of only verbally describing a work of architecture or construction process. For Mario Carpo: 'Alberti … tries in the De re Aedificatoria to emulate through plain alphabetic writing the expressive potential of the images whose use he rejected.'[9]

De re Aedificatoria is perhaps one of the first examples of the early Renaissance desire to both establish architecture as a profession and disseminate information about its practice – a task made relatively easy by Gutenberg's press.[10] This is in direct opposition to the operations of the late Gothic guilds, from which Brunelleschi was born, which sought to keep construction practices as highly guarded secrets and translate them mostly verbally.

In writing his treatise, Alberti became an advocate of those architects engaged in design but not in construction. This codified a split, or gap, between design and making that still exists today, more than 600 years later, and frequently puts architects at odds with those who build their work. Alberti's 'disinterest in the actual realization of his designs may have been a consequence of the forma mentis (mindset) he acquired in the cautious, reserved circles of the Curia. Or it may have been the result of a natural preference for the purely theoretical aspect of his art.'[11] In fact, this tendency towards broader intellectualism, as opposed to trade specialisation, 'comes to the fore again and leads to the cult of a type of versatility which is more akin to the dilettante than the craftsman'.[12] This is not to suggest that Alberti's vast contributions, both theoretical and pragmatic, are not significant to the contemporary practice of architecture; rather, that the framework within which he sought to position that practice of architecture was rooted in the discourse of making, and not in making itself. 'It is evident from the disdain with which Alberti refers to building masters that he imagined them capable only of execution and not of conception in architecture.'[13]

Interestingly, though Alberti most famously advocated this separation of the design profession from the building trade, documentation exists that suggests this division already existed. By the early 15th century, late medieval Gothic builders in Europe were, unbeknown to them, already working according to what has come to be known today as the Albertian Paradigm.[14] As early as the mid-14th century, construction officials were formally adopting the title 'architect' or 'archititector', and working remotely – and even internationally.[15]

Alberti and the Advocation of Models

In the second book of De re Aedificatoria, subtitled Materials, Alberti discusses the use of models in architectural design. Like Brunelleschi, he also suggests that models should remain incomplete, proposing that 'the presentation of models that have been colored and lewdly dressed with the allurement of painting is the mark of no architect intent on conveying the facts; rather it is that of a conceited one, striving to attract and seduce the eye of the beholder, and to divert his attention from a proper examination of the parts to be considered, toward the admiration of himself'.[16] What is remarkable about this section of the work is how Alberti articulates the use of models, and how these uses are similar to the purpose of BIM systems in contemporary practice. This is an interesting contradiction: Alberti's call for the separation of design and making would seem consistent with architects making representations of their designs with models (for example, renderings), but instead he recommends more operative uses.

Alberti suggests that the use of models to examine, in an iterated way, the relationship of a design proposal to its site and district, its overall form as well as the internal relationships between its components, is paramount to understanding how appropriate a proposal is. Further, the adequate size and shape of such components leads to a proper selection of materials and orders, and quantity of columns: 'their thickness … extent, form, appearance, and quality, according to their importance and the workmanship they require.'[17] This suggests the model can be used to investigate construction sequencing and techniques as well as costs of materials, and that the number of elements can be counted so that a budget can be arrived at in a similar way to how Brunelleschi studied the vaulting of the dome of Santa Maria del Fiore through his models and inventions, including the wood (and possibly stone) chains that lent tensile support. Likewise, the digital information models of today simulate construction sequencing over time to allow for a better coordination of trades, and are linked to live databases to calculate real-time construction estimates.

Differences Between Alberti's and Brunelleschi's Uses of Models

While both Alberti and Brunelleschi employed models in their advocation or execution of architectural design, there exist between them telling differences that speak to their own understanding of architectural practice and are of interest for discussions about BIM today. For Alberti, an architectural idea, or disegno, is conceived in the mind, but is only realised through a model.[18] The model is a mechanism through which an architectural idea unfolds. Alberti is less concerned with the model's ability to convince a client of the appropriateness of the proposal; whether it is appropriate or not is determined through the intellectual development of the proposal itself. He makes the clear distinction that a model is used firstly to fulfil the creative dimension of a design proposal, and secondly in the pragmatics of construction.

Brunelleschi thought that a model was a representation of an idea already formed in the mind, hence his secrecy about fully disclosing his intentions to others, especially workmen on site. For him, the model served as a virtual construction of an actual building. How specifically

Brunelleschi's ideas were formed prior to the construction of the models we will never know. However, it is clear that the making of models for the various parts of the Santa Maria del Fiore dome (for example, its lantern and scaffolding) and some of his inventions were critical to the successful completion of that project. In contrast to Alberti, for Brunelleschi intellectual development and construction process or method were comprehensively integrated in the model.

Both of these positions seem to overlap and contrast with contemporary discourse in the adoption of BIM as a new paradigm for the design process. Through simulation and the ability to form/fabricate materials directly from digital models, writers such as Sanford Kwinter and Manuel DeLanda have posited that information models have allowed us to enter a new paradigm: 'the virtual to actual'. Here, a building is already fully real and simply needs to be actualised via translation from virtual to actual matter. This is contrasted with the 'possible to real', a paradigm advocated by Alberti that has existed for the last 600 years. In this approach, the possible (an idea) has no measured relationship to the real because it is necessarily interpreted by a third party – in the case of construction, a builder who interprets a set of architectural drawings (representations). DeLanda writes more specifically that the former is a design philosophy that takes into account materials (both virtual materials simulated within the computer, and actual materials used in building), while the latter is purely cerebral and as such has no relationship to matter or materials.[19]

Though there is no reference to either of these paradigms in the work of Alberti or Brunelleschi, they are helpful in tracing the impact of their ideas about the practice of architecture. For both, there is an important distinction in the utility of models as devices to work through the problems of construction. In advocating the separation of design from making, Alberti's use of the model in this capacity was to arrive at a subjective appropriateness of a design proposal. Brunelleschi, his secretiveness and desire to control the site notwithstanding, used the model as a collaborative apparatus to virtually work through the problems of construction – ideally with those who would be responsible for the actual construction. He was known to carve details for his workmen in wood, clay and wax. This notion of enhanced collaboration, as opposed to separation, forms the essence of information modelling technologies.

Closing the Gap *Due*

It is in fact impractical to suggest that the architect's return to the role of master builder would discourage those architects working remotely today. Although it is clear that the success of such a mode of practice was in part due to the proximity of the designer-builder to the building site, and access to workers and indigenous building materials (such as the use of Carrara marble from in and around Florence for the Santa Maria del Fiore dome), the ubiquitous nature of information has allowed for a change in the architectural delivery process. In the Albertian Paradigm, interpretation was necessary to mediate between an architect's intentions and the realisation of the building – a 'possible to real' relationship that could not ensure precision in the translation from drawing to building.

Certainly, one of the advantages of information models being advocated here is that they have the ability to translate and actualise data from the virtual state. In both the pre-Renaissance and immediate contemporary condition, the collaborative aspects of this notion – that the designer/architect and builder/tradesperson could physically work on a scaled construct testing ideas, building techniques and construction sequences – seem to indicate that the gap between design and building, conceptualised by Alberti and which existed throughout the 20th century and into the 21st century, will finally be closed. ⚙

Notes
1. See Ross King, *Brunelleschi's Dome*, Penguin Books (New York), 2000.
2. See Frank D Prager, 'Brunelleschi's Inventions and the "Renewal of Roman Masonry Work"', *Osiris*, Vol 9, March 1950.
3. These wardens were themselves members of Florence's largest and most powerful guild – the Wool Merchants.
4. Christoph Luitpold Frommel, 'Reflections on the early architectural drawings', in Henry Millon and Vittorio Lampugnani (eds), *The Renaissance from Brunelleschi to Michelangelo*, Bompiani (Milan), 1994, p 102.
5. Henry Millon, 'Models in Renaissance architecture', in Millon and Lampugnani, op cit, p 22.
6. Ibid, p 72.
7. See Prager, op cit, p 509.
8. Leon Battista Alberti, *On Painting*, revised edition, Yale University Press (New Haven, CT), 1966, p 40.
9. Mario Carpo, *Architecture in the Age of Printing*, MIT Press (Cambridge, MA), 2001, p 124.
10. See ibid, p 9, where Carpo suggests that the 'modern print format' afforded by the printing press affected the transmission of architectural theory in the 1530s.
11. Franco Borsi, 'The themes of Alberti's life', *Leon Battista Alberti: The Complete Works*, Rizzoli (New York), 1986, pp 10–11.
12. Arnold Hauser, 'The Social Status of the Renaissance Artist', in *The Social History of Art*, Vol 1, Knopf (New York), 1952, p 335.
13. Franklin Toker, 'Gothic Architecture by Remote Control: An Illustrated Building Contract of 1340', *The Art Bulletin*, March 1985, p 88.
14. This term should be credited to Mario Carpo, who has used it in several texts including 'Nonstandard morality: Digital technology and its discontents', in Anthony Vidler (ed), *Architecture Between Spectacle and Use*, Yale University Press (New Haven, CT and London), 2008, p 131.
15. See Toker, op cit, pp 67–9.
16. Leon Battista Alberti, *On the Art of Building in Ten Books*, trans Joseph Rykwert, Neil Leach and Robert Tavernor, MIT Press (Cambridge, MA), 1988, p 34.
17. Ibid.
18. Henry Millon, op cit, p 24.
19. Manuel DeLanda, 'Philosophies of Design: The Case of Modeling Software', *Verb: Processing*, Vol 1, No 1, 2001, pp 132–42.

Recycled Toy Furniture

2008

Greg Lynn FORM

By Greg Lynn

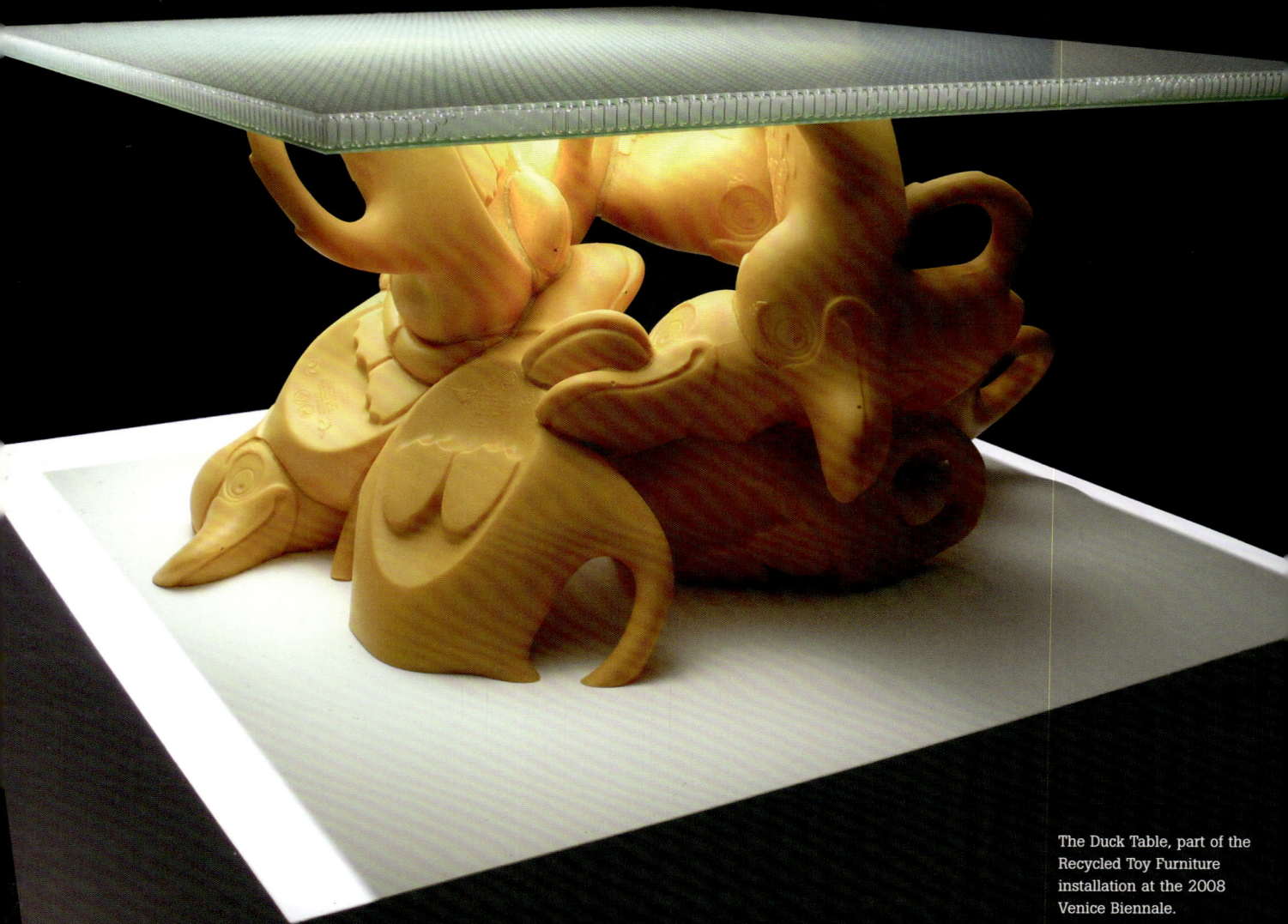

The Duck Table, part of the Recycled Toy Furniture installation at the 2008 Venice Biennale.

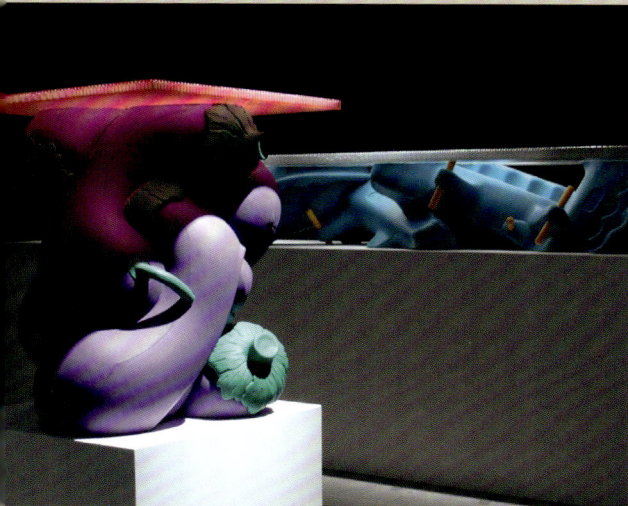

The Eggplant Table and Whale Table, two pieces exhibited as part of Lynn's Recycled Toy Furniture installation at the 2008 Venice Biennale. Lynn was awarded the Golden Lion for Best Installation Project at the international exhibition. He has recently put out an internationl call for various types of toys on his website, where he offers to buy used toys to convert into the designs seen here. His recycled toy constructions are rustic, curvaceous, globular, moulded and playful: they are toys after all.

These first-generation prototypes of the high-tech scavenging of recycled plastics for furniture include four different-sized tables with plastic Panelite tops, a low bench, storage wall, coat rack and cylindrical shoe closet.

Our everyday lives are surrounded by plastic water bottles, plastic cars painted to look like metal, plastic furniture, plastic implants in our bodies, plastic additives in our concrete, plastic wall materials, light filtered through plastic diffusers and thin plastic screens on our desktops. It is therefore no wonder that glass seems so exotic to architecture when we are surrounded by so much plastic.

I have been preoccupied with recycling my own children's toys into walls, furniture and usable objects; using their toys as bricks. This led to the design of the Blobwall system of construction (see overleaf), the first plastic brick that brings everyday life into a building-scale masonry construction system. The firm has now found a way to scan, design, fabricate and construct walls of recycled toys that are lightweight and bear their own weight. In addition, they do not use the labour or expertise of masonry and the wet forgiving technology of mortar to become level and true. Instead, they are laser-scanned and digitised into a computer, designed and arrayed like bricks, their intersections are defined as cutting paths, and a robot cuts their joints and connections with precision. They do not rest on mortar joints, and are not even glued; the toy bricks are welded together with a tool used to repair car fenders.

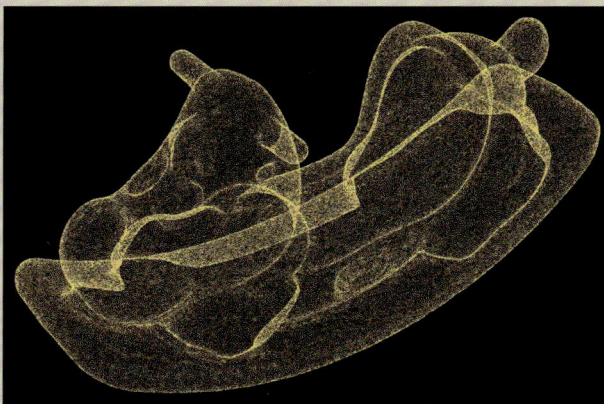

Computer image of the 3-D model of a dog toy that was 3-D laser-scanned into the computer. The toys are translated into the computer at Reverse Modeling, Inc, using a structured light scanner otherwise known as a 'white light' scanner. The scanner produces a dot pattern on the object and translates the information into 3-D points. Each scan view (top, bottom, sides) is combined to create the overall 3-D model as point clouds, which are automatically translated into meshes by the computer software.

A six-axis rotational moulding robot arm cutting a duck piece. The rotational moulding machine uses heat and biaxial rotation to cut the toy pieces, which are then welded together with a hot welding system from Machineous.

Screen capture of the Duck Table, displaying how the scanned toys are rotated and intersected in Maya.

Blobwall

2006-08

Greg Lynn FORM

By Greg Lynn

The blob unit, or 'brick', is a tri-lobed hollow shape that is mass-produced through rotational moulding. Each wall is assembled from individual robotically cut hollow bricks that interlock with exact precision.

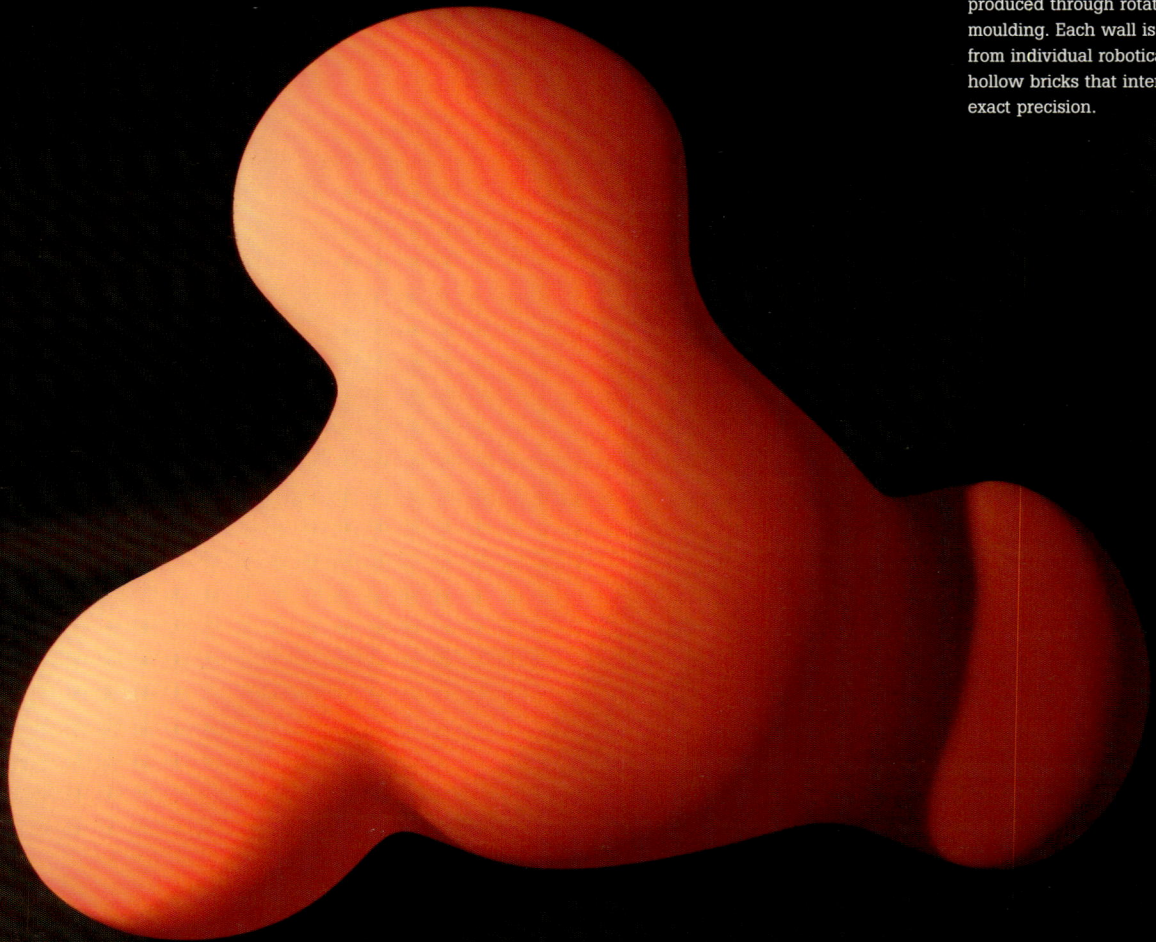

The design of Blobwall begins with a redefinition of architecture's most basic building unit – the brick – in lightweight, plastic, colourful, modular elements custom-shaped using the latest CNC technology. The freestanding, indoor/outdoor wall system is built of a low-density, recyclable, impact-resistant polymer. The blob unit, or 'brick', is a tri-lobed hollow shape that is mass-produced through rotational moulding. The stock designs are: S-, L-, I- and U-shaped walls as well as a Dome and Tree House. Each wall is assembled from individual robotically cut hollow bricks that interlock with exact precision.

Each of the colour schemes is made up of a gradient blend of seven different brick colours. In the Renaissance, palaces were designed to mix the opulent and the basic, the elegant and the rustic. Stones were hewn so that planar faces could stack and bond, but the outward faces of the stones formed cloven and rustic facades. The Blobwall recovers the voluptuous shapes, chiaroscuro and grotto-like masonry textures of baroque and Renaissance architecture in pixilated gradients of vivid colour. It is both product – like a child's toy – and building.

The Blobwall at 'Greg Lynn FORM: Blobwall Pavilion', an exhibition at SCI-Arc in Los Angeles which ran from May to July 2008. For this installation, Blobwall units – or 'brick's – were assembled as the Blobwall Pavilion, a hollow, illuminated structure. Each of the bricks is fitted with a tiny computer-controlled light and at night the pavilion comes to life, taking advantage of its pixilated masonry construction.

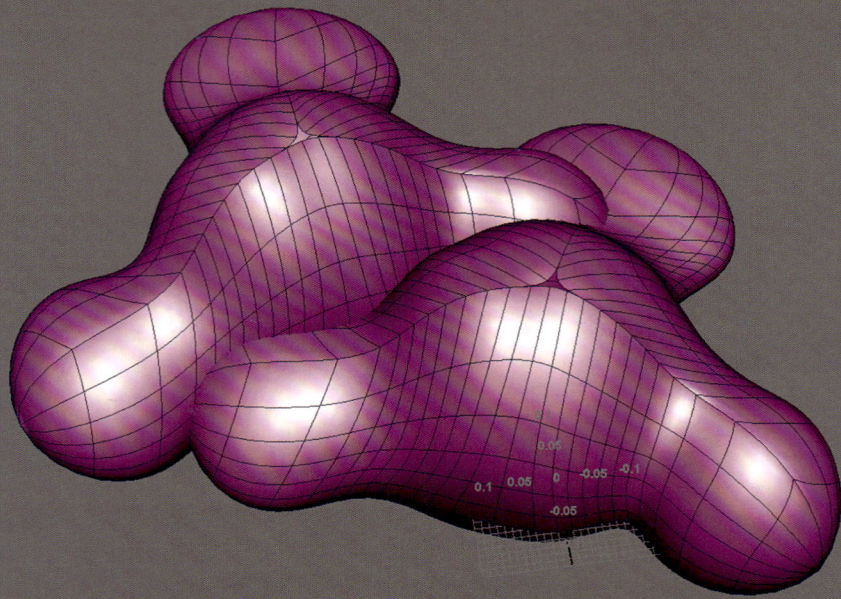

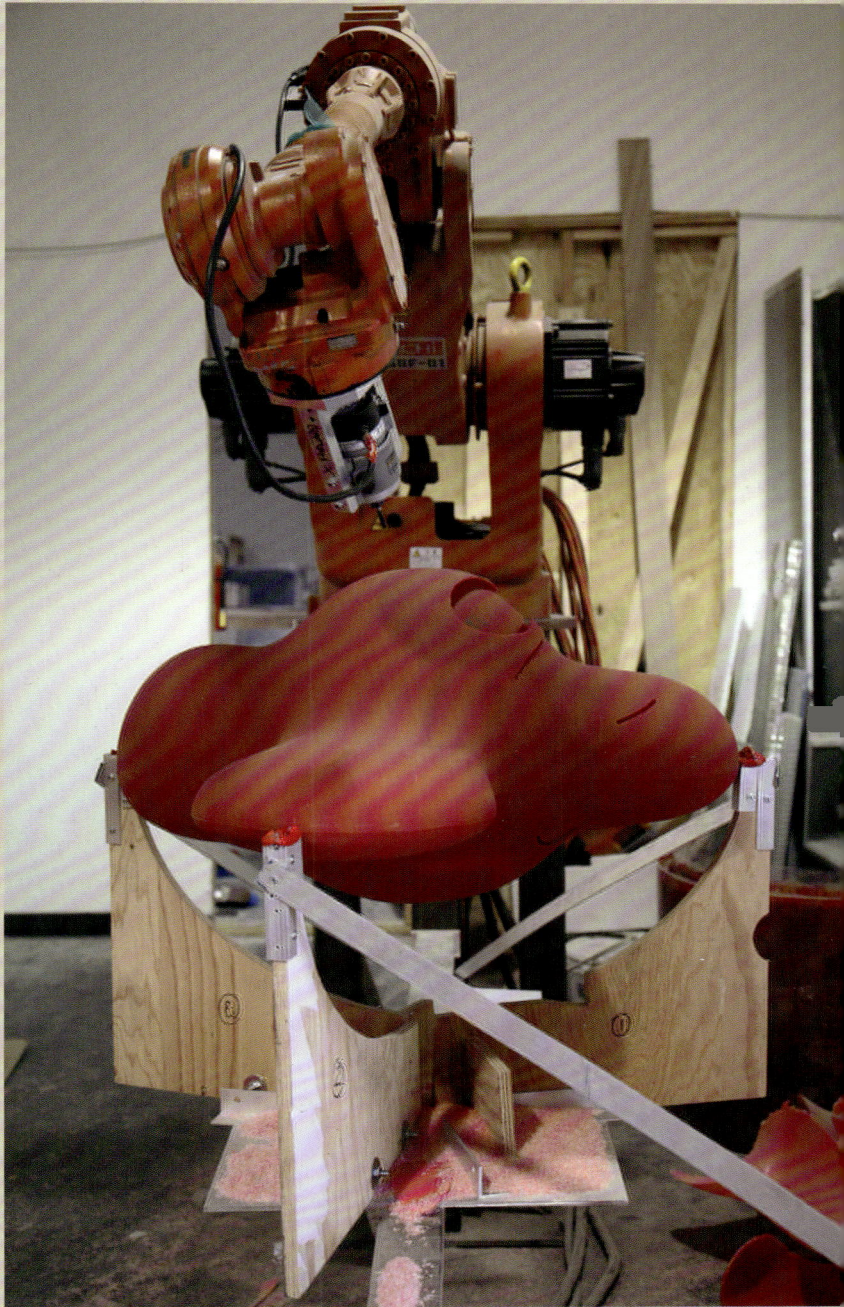

above: The Blobwall in blue. In this contemporary rusticated wall, the three-lobed form of the bricks means they can tuck together nose to forked tails and also, when rotated in a gradient series, can become more lumpy and articulated.

right: The brick being cut by a rotational moulding machine.

opposite top: Screen capture of the blob unit, outlining the tri-lobed hollow shape. Here, two blob units are being intersected in Maya.

opposite bottom: The Blobwall at the 'Skin + Bones: Parallel Practices in Fashion and Architecture' exhibition at the Museum of Contemporary Art in Los Angeles, which ran from November 2006 to March 2007. ∆○

Our everyday lives are surrounded by plastic water bottles, plastic cars painted to look like metal, plastic furniture, plastic implants in our bodies, plastic additives in our concrete, plastic wall materials, light filtered through plastic diffusers and thin plastic screens on our desktops. It is therefore no wonder that glass seems so exotic to architecture when we are surrounded by so much plastic.

An Enthusiastic Sceptic

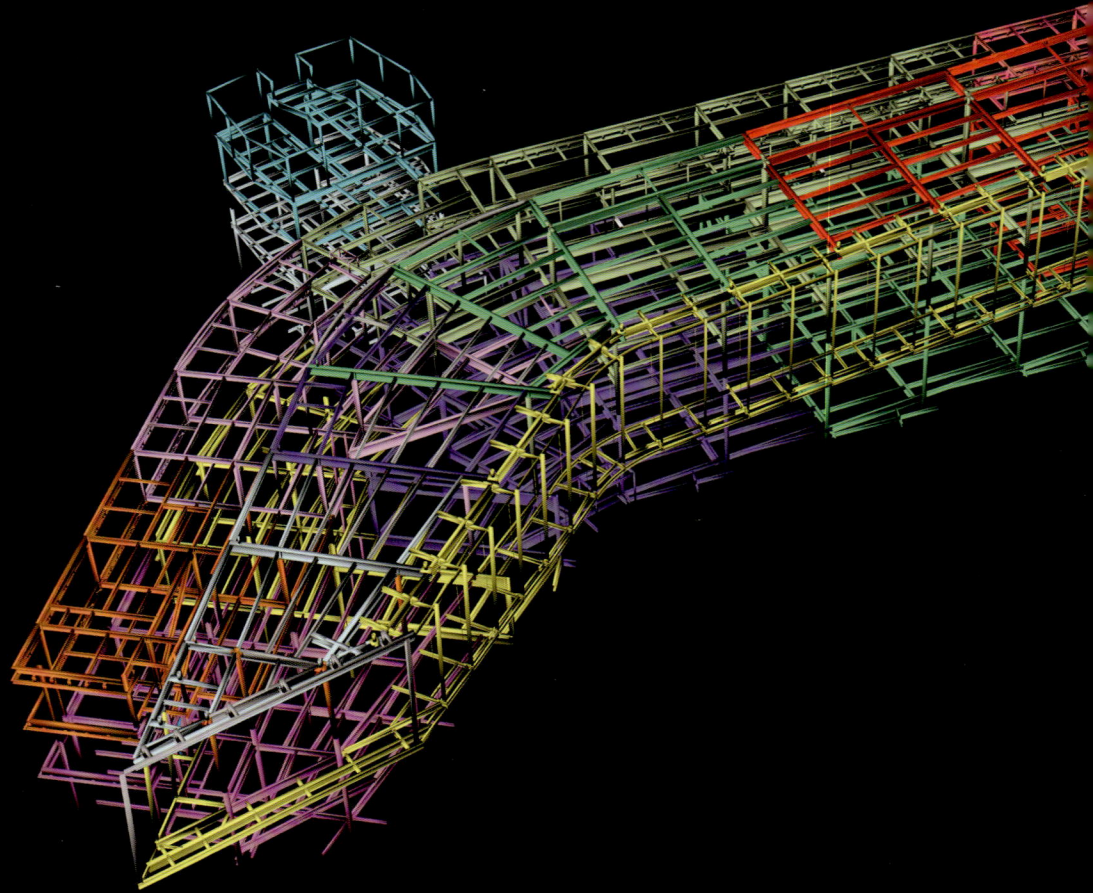

Nat Oppenheimer, principal of structural engineers Robert Silman Associates, is a dissenting voice among the advocates of building information modelling (BIM) – an enthusiastic sceptic. Could the quest for integration be leading towards oversimplification rather than customisation and differentiation? Even towards the sort of standardisation adopted with LEED ratings in the quest for sustainability?

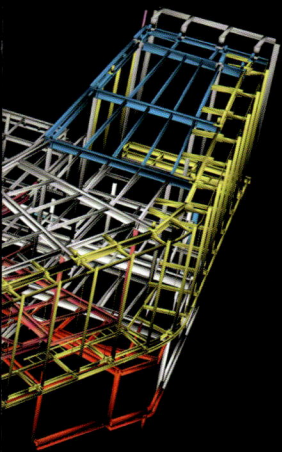

The sceptic, if he or she does his or her job correctly, is critical to the productive evolution of any movement, and as we embark on the Building Information Modelling (BIM) revolution, it is worth asking a few questions.

We embrace the promise of BIM – born from the notion of collaboration between disciplines and the integration of design – and recognise it as a paradigm shift in an industry that has spent too much of the past century trying to separate consultants on the same design team, to clarify liability and to shed responsibility. Any move towards accepting more responsibility is generally a good direction, and we are therefore enthusiastic about the potential of BIM and the cultural shift it appears to be initiating.

Nevertheless, we are concerned that the demands of the market will dampen the original promise and leave us, 20 years hence, with nothing more than an advanced pen-and-pencil set rather than a truly different way of designing and building iconic architecture.

Any discussion about BIM must recognise the broad field of architecture, engineering and construction, and note that any movement that tries to capture the imagination of the entire industry will never be able to be everything to everyone. This view is built on experience within the small corner of the industry that focuses primarily on unique, one-off designs that often aim to break convention rather than embrace it. Over the last 20 years, the structural engineers at Robert Silman Associates have continually sought out such projects, which have required them to overcome the challenges they present from many different angles. Software and industry shifts towards any sort of 'regularisation' or common denominator design thus hold little interest for the firm.

While BIM is alive and well in this field – and thriving in many academic circles and smaller firms – the current marketing trend for the industry does not seem to embrace many of the directions these smaller firms explore (custom fabrication, odd geometric shapes, custom mechanical systems and so on). This is unfortunate, as some of the most interesting advances are generated from their research. In addition, the move towards integration appears to be leading to oversimplified buildings, lower fees and shorter design schedules rather than to the quest for the perfect jewel of a project. It is therefore necessary to question what this revolution strives to accomplish.

Does a small and inventive firm (or a large firm that thrives on model building and sketching to generate magic) risk losing its mojo by relying on a software platform designed for the larger sector of the industry? Does the drumbeat towards more standardised BIM software overwhelm some of the more exciting opportunities for integrated design being promoted in small firms and academia? Can the promise of very advanced proprietary BIM software developed by large architectural and engineering firms be harnessed by generic software for use across the whole industry? And will the industry embrace the liability and responsibility inherent in the world of BIM?

BIM and High-Performance Design

In asking these questions and looking at alternative futures – as opposed to the shiny ones most often promoted within the BIM community – high-performance (green) design offers a cautionary tale from which to learn. Analogous to BIM, the heart and soul of successful green design depends on the collaborative and integrated effort of the entire design team. However, the definition of 'success' in green design (LEED excellence versus carbon neutral, etc) has remained a moving target within the industry. In the same way, one can ask what the true definition of success in BIM is. Is it simply a more streamlined design and production effort with fewer conflicts between architecture, structure and mechanical systems, or is it a truly integrated, unique building where concept, drawing production, fabrication and, ultimately, building management systems are all tightly integrated, with primary control held by the design team?

At this point in the history of high-performance design, the market looks to the LEED credit worksheet as a way to control the chaos inherent in any design and, while there is great value in these worksheets, they frequently dumb down the process and turn the lofty goal of green design into one of bean counting and horse trading. LEED has recognised that it cannot be everything to everyone and has begun to separate its worksheets into different categories. But there is still the overall simplification of design that has at times hurt the overall advancement of new technologies and real change within the design of buildings and the way energy is used.

The Prehistory of CAD and BIM

With the rise of CAD came the power to harness ink and mylar into a much more precise version of the same drawing. This moved the field of document production into the digital age in a two-dimensional way, by simply bringing the standard drawing alive as a series of lines and symbols. The proof that the shift to CAD was just a digital adaptation of the old industry standard lies in the fact that many engineering offices still have a drafting department (retrained in CAD) and still separate the task of engineering from the task of drawing.

BIM, by its very nature, gives life to the lines and symbols on the sheet and works to reintroduce the engineer to drafting, and the drafter to engineering (traditional drafters were able to put together many components of a standard set of drawings without input from engineers – a skill that has disappeared over the last 20 years with the loss of

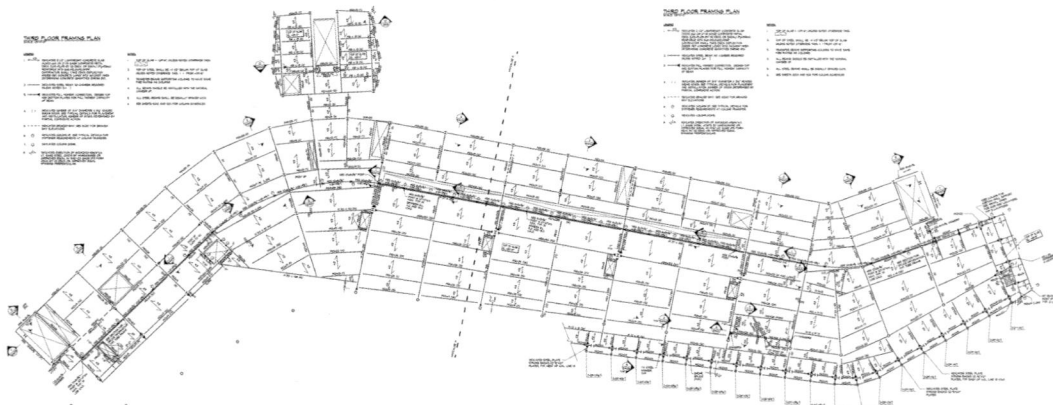

Robert Silman Associates, Lewis Katz Building, Dickinson School of Law, Pennsylvania State University, University Park, Pennsylvania, 2006
Framing plans, first drawn by hand in pen, of the Whitney Museum in New York, and a current CAD document of the law school building. Note the similarity between the two documents. The efficiencies of computer-aided drafting aside, very little has changed in the basic transfer of information even though the manner in which each document was produced differs. Both are notational drawings and as such need to be interpreted by a general contractor and structural steel fabricator.

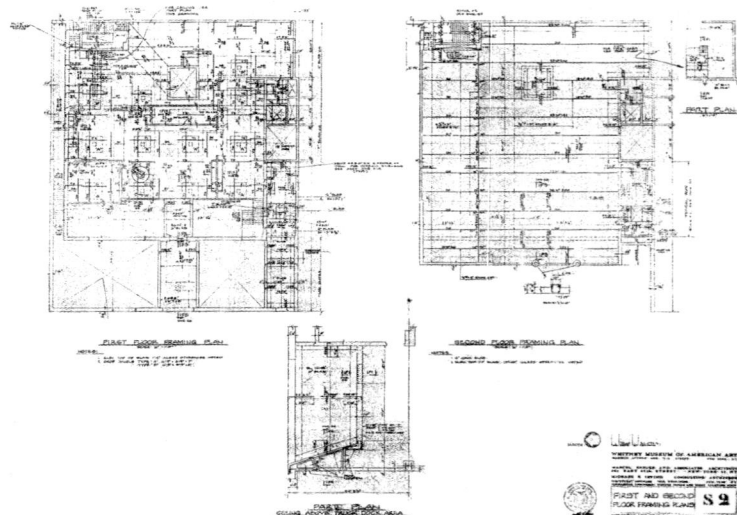

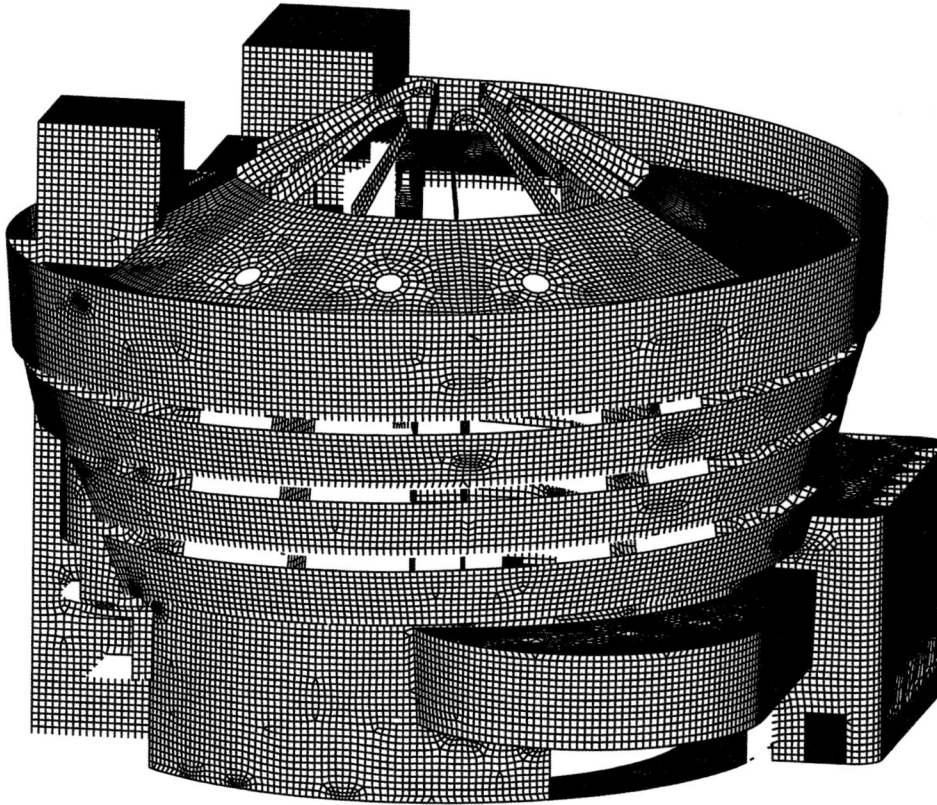

technically trained drafters). This is a very valuable collaboration tool within any office. However, the sceptic sees the current stage of BIM, in its most common usage, as just a more powerful version of CAD where it is primarily used to coordinate all trades within a design but not necessarily to create. Does this suggest that purveyors of custom architecture, therefore, should not waste their time with the current versions of BIM software? Far from it: as with CAD, it was the proficiency of the industry and the widespread acceptance that eventually made the true shifts in the construction documentation process possible. Proficiency in the use of BIM, therefore, will likely lead to more creative uses of the platform in the future.

In examining the current state of BIM, it is interesting to remember that CAD came out of CAD/CAM and the promise of the full automation of factory fabrication. The software originally developed to make perfect pieces of machinery was co-opted by the architectural industry essentially to make beautiful drawings. Will BIM be the paradigm shift that brings architectural drawings to life by moving seamlessly from concept to integration to fabrication, or will it fall apart in a wave of liability fears and take root as yet another tool to make better documents instead of better projects?

Atypical BIM Usage

As a sample of the atypical uses of BIM at Robert Silman Associates, the following three case studies offer a glimpse into the successful application of integrated design and project delivery. Each project gives a sense of the potential of BIM beyond the perfect set of documents and how much background work is behind any successful project. In addition, all demonstrate the human energy that still underlies the success of any integrated design.

Managing Existing Building Information: The Guggenheim Museum

While much has been said about the power of BIM to coordinate trades within a new building, less has been made of the power to re-create existing conditions and the ability to then analyse existing structures with more precision.

In 2005, Robert Silman Associates was asked to solve the issue of recurring cracks in the facade of the Guggenheim Museum in New York City. The engineers were tasked with determining what was causing the cracks to reoccur after each previous repair campaign and, if it was structural in nature, to assist the restorers on the project in selecting the appropriate material to be used as a filler and coating. To do this, it was first necessary to create a full-scale, as-built structural model of the Guggenheim – a monolithic, non-orthogonal, concrete structure.

Working with digital survey company, Quantapoint, Robert Silman was able to create the entire structure of the Guggenheim's main

Photograph of an interior radial monitor within a structural cavity of the Guggenheim Museum. A series of string potentiometers mounted on custom brackets was installed to measure movement at the top of the wall. The data was used to generate a mesh information model in which a series of small structural cracks in the structure could be studied. The model could be further stressed, virtually, to predict how the cracking would impact the concrete structure over time.

rotunda in SAP, a finite element modelling software, and model this as a continuous shell. Taking field data of actual movements over the course of a year, the model could be calibrated to events in the field so that the future movements of the building and the impact of any of the design proposals could be predicted.

In the end it took six months of painstaking, analogue-type work (for example, checking the meshing of each node among many millions) to establish the cause of the movement. However, each successive step in the evolution of all of the software used in the Guggenheim project will simplify future efforts, and the experience gained will lead to much better interpretation of the results achieved in future versions of the software.

While it is recognised that this particular use of digital information management and data transfer is quite unique, it is hoped that in future it will be possible to import an existing building into finite element software for full-scale analysis. This will allow such buildings to be looked at in a new light – as an assembly instead of individual elements (which, more often than not, do not meet contemporary code requirements even though the building as a whole has stood without incident for many years) – and perhaps enable us to reach a better understanding of the overall capacities of structures in their entirety.

Front End Versus Back End: The Lewis Katz Building, Pennsylvania State University

While the ultimate integration of design documents and shop drawing within a single BIM platform is already in use, it is the use of BIM by subcontractors to produce well-coordinated and thorough shop drawings that is the more rapidly evolving industry standard. With the design

established and the liability for the fabrication strictly in the subcontractors' hands, they can take a set of fully developed drawings and virtually construct the building before erecting it physically.

To date, the Lewis Katz Building of the Dickinson School of Law at Pennsylvania State University has been one of Robert Silman's most successful projects involving BIM in this way. The project is a large, complex design with numerous design features and an aggressive schedule. As a testament to the rapid integration of BIM, the design team had not used BIM on the project because it was not fully integrated in their offices when the work commenced two years hence.

For this project, the steel fabricator used BIM software to develop the model of the complete steel frame for production of the shop drawings. In the very thorough question-and-answer stage of their modelling, steel coordination issues were worked through in an efficient and productive way.

This use of BIM is, in itself, establishing a new level of clarity in the construction of buildings. For those members of the design and construction team who had worked in this manner before, receiving a significant number of requests for information (RFI) during the building of the BIM model was not unusual, and meant, ultimately, that when set up correctly, the building frame would go up without incident or a single RFI once the steel was on site. Though everyone on the project recognised that this was the preferred sequence, it still gave pause to those not familiar with the power of BIM when there was a tenfold increase in RFI at the very beginning of the project.

While it goes without saying that full integration of BIM is intended to result in a seamless transition from design to erection, there are concerns that the loss of human communication between the design and construction phase will have unintended consequences on the collaborative process. If we rely too heavily on the black-box solution, we risk losing the vital education that often occurs between design and construction. Recognising that, at some point, this will evolve into a wonderful place where the integration software is robust and allows us

Peter Gluck and Partners, Winnetka House, Illinois, 2006–09
Axonometric drawing by Peter Gluck and Partners detailing the layout of the Winnetka House's poured concrete foundation. The document was generated from an information modelling system that automates dimensioning and is used for bidding. The drawing was transmitted to the foundation subcontractors to reduce the concerns that may occur when an atypical foundation is simply noted in plan and section. As contracting firms further embrace the utility of BIM it will become possible to transmit digital files directly to them for the automated manufacture of concrete formwork and the generation of shop drawings. In addition to the perimeter condition, a series of internal steel columns to transfer point loads are noted.

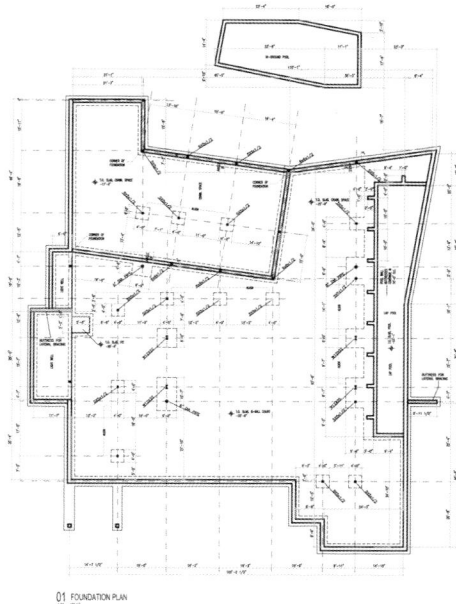

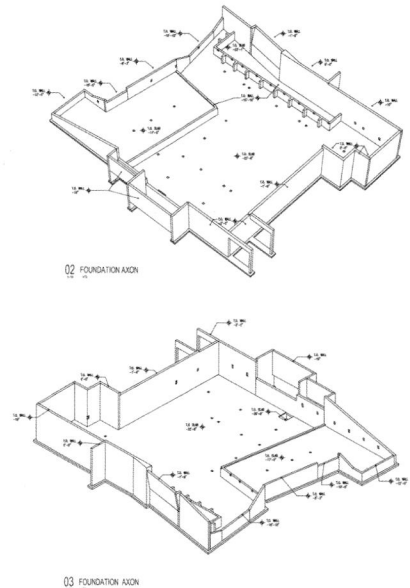

to concentrate on other things, we should tread lightly over the next few years of working through the bugs to get there.

Peter Gluck and Partners

Robert Silman Associates has worked with New York architects Peter Gluck and Partners for the past 10 years in a way that is unique to the industry. The collaboration is essentially an analogue version of BIM, in that each staff architect is expected to understand the impact of every component of the design, including all structural and mechanical elements. As an extension of this, the office produces all of the documents for the project, with consultants marking up the architectural documents with structural information.

More importantly, Peter Gluck is a true design-build firm, where each staff architect is expected, at some point, to manage the construction of a project and all of the subcontractors. This gives the architects a unique perspective on the importance of coordinated drawings and the management of information. Interestingly, the firm is not yet using BIM to manage its projects. It has achieved its best results by releasing stand-alone structural packages (integrating MEP and architecture) bit by bit as the site needs them, and feels that while each package is highly integrated, BIM, at this time, may bog down this system by looking for an overall integrated solution that is often still in the designers' minds when the early sets of drawings are released to the site.

This demonstrates the current paradox of BIM: with only a small percentage of the industry truly facile at using the software, an integrated firm such as Peter Gluck and Partners with a tremendous reliance on direct hands-on input for all of its projects finds that the

software is, in some ways, not facile enough for its needs. In this example, the firm is a very effective living version of BIM, and when it does decide to move into producing project documents with BIM software it will be in a position to control (because it knows the integration of buildings so well) instead of being controlled by the software and told how to integrate based on the algorithm within the code.

Conclusion

The examples here have tried to look beyond the more standard uses and preferred rhetoric of BIM. The basis of Robert Silman Associates' mission is consulting on unique projects that are rarely a repeat of one that has come before. This often means adopting software that is more tailored to the orthogonal and the repetitive, and adapting it to the uniqueness of each project. Experience in this field frequently compels the firm to question the potential of the first version of any software and immediately look at how it can be bent, twisted and adopted elsewhere.

So far, there is nothing to suggest that BIM software does not, and will not, successfully address the issue of uniqueness. And there is no reason why BIM cannot develop along many different paths. However, human nature being what it is, it is still possible that the industry will see only the simplifying virtue of BIM software and thus allow it to define the design rather than the other way round.

As certain parts of the industry awaken to the 'bean counting' downside of LEED spreadsheets and demand more from green design, we are confident that there will be some who will always push this software to do more, and that these pioneers will encourage clients not to accept the simplistic and seemingly perfect in lieu of the truly integrated. △D

Toni Stabile Student Center
Columbia University Graduate School of Journalism

New York, 2008

Marble Fairbanks

By Scott Marble and Karen Fairbanks

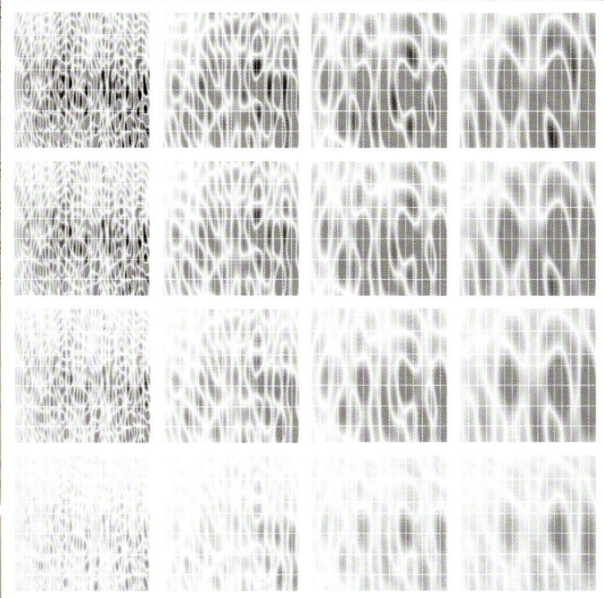

The ceiling of the social hub, comprised of perforated 16-gauge steel panels backed with acoustic insulation, provides an acoustically absorptive ceiling for a room that accommodates a wide range of uses that require sound control. The geometry of the ceiling wraps tightly to the existing Journalism School building to increase ceiling height where possible. The perforation pattern was developed through a two-phase process: first, an acoustic model of the space was developed to drive the density of perforations, and a second subsequent scripting process integrated geometry, lighting and sprinkler layout within the pattern generation.

The Toni Stabile Student Center project consists of a partial renovation and addition of a glass-enclosed café to the existing McKim, Mead & White Journalism School building on the Columbia University campus. The renovation includes a new social hub for students, flanked by the journalism library, faculty offices, classrooms and a student newsroom. The focus of the project was the design and fabrication of performance-driven surfaces using quantitative criteria taken from digital analysis models. The surface types included acoustic, graphic, solar and mechanical, and the criteria adopted for designing and engineering each surface were developed directly from the technical and programmatic demands of the space in which it was located.

In addition to the main design consultants, a team of collaborators with expertise in each of the performance types was assembled to contribute to developing the specific criteria and design for each surface. Working closely with fabricators, each surface system underwent rigorous full-scale physical prototyping. While the technical performance was numerically driven and calculated through digital simulations, the qualitative effects of each surface were tested at full scale to confirm the desired resolution, legibility and overall effect. The techniques of the assembly for each surface were finalised in the prototyping phase with information and coordination of the assembly logics incorporated into the final digital fabrication files.

An acoustic model of the social hub was developed to establish the performance criteria for the ceiling's perforation pattern. Several scenarios were generated within the software to identify the zones of the ceiling that, through increased acoustic transparency, would reduce and eliminate the effects of reverberation in the space. These points then became 'zones of intensity', or 'attractors', for the pattern-generation script – areas where the perforations would become larger and provide more acoustic absorption. The pattern script was developed to generate a series of unique iterations, each of which relied on the attractor points and thus satisfied the acoustic performance criteria. The iterations were evaluated both for the density of perforations (which translates directly to fabrication time and cost) as well as overall qualitative effect.

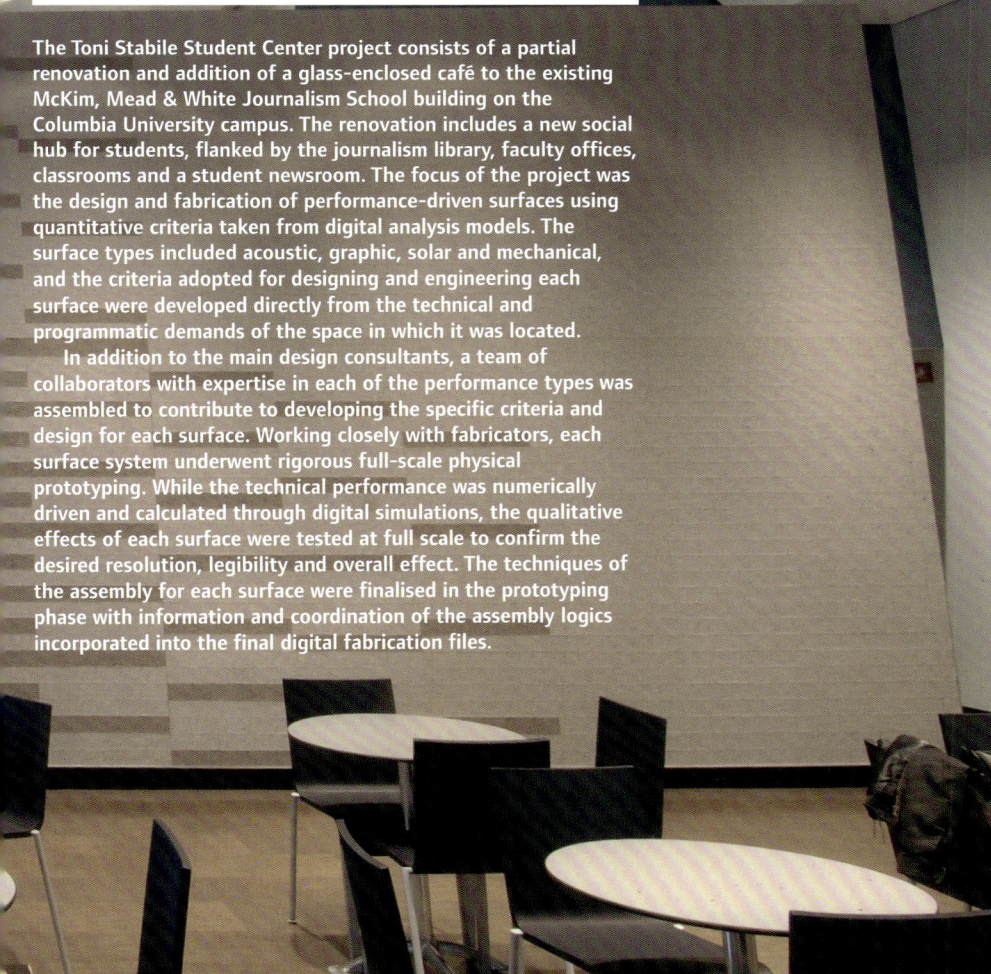

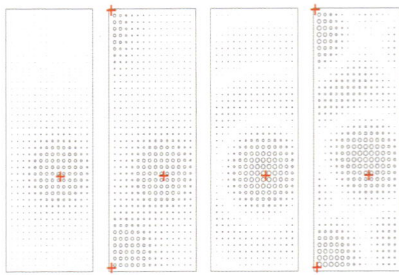
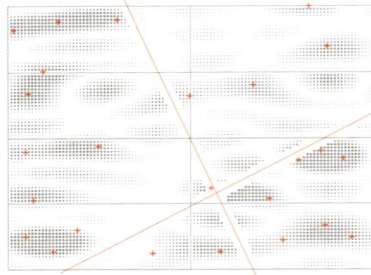
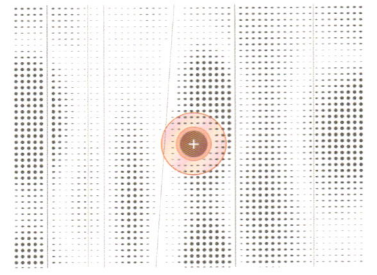

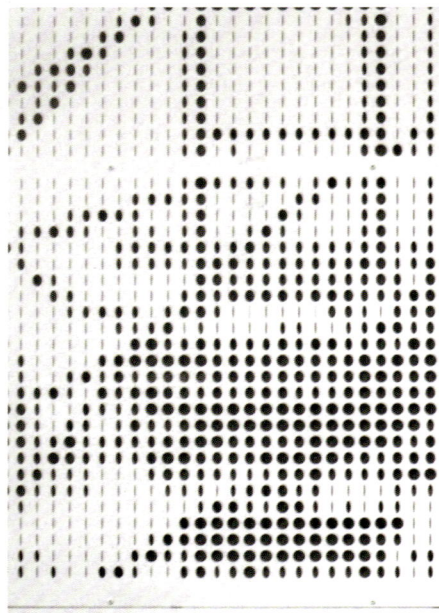

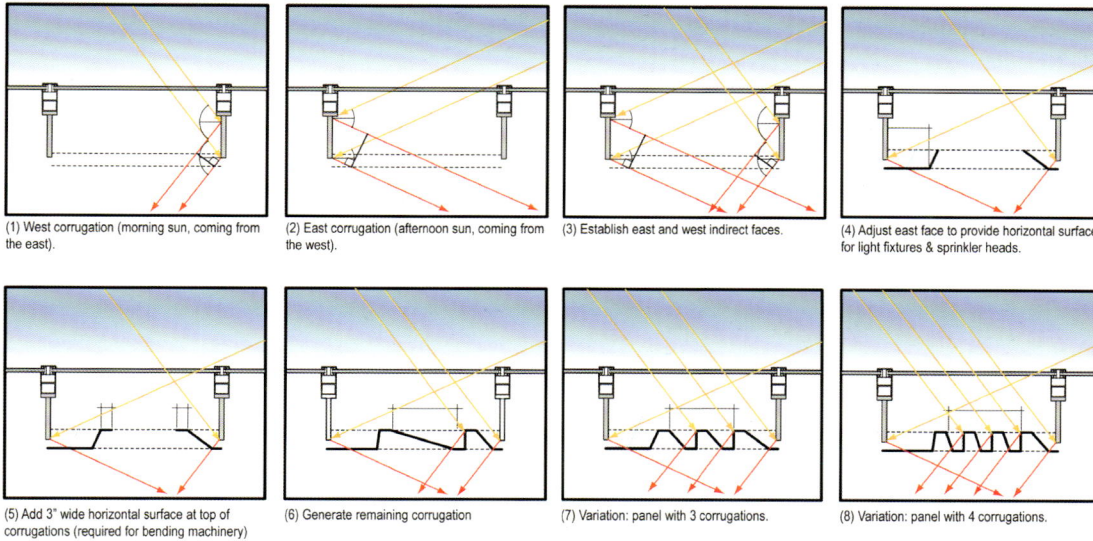

(1) West corrugation (morning sun, coming from the east).

(2) East corrugation (afternoon sun, coming from the west).

(3) Establish east and west indirect faces.

(4) Adjust east face to provide horizontal surface for light fixtures & sprinkler heads.

(5) Add 3" wide horizontal surface at top of corrugations (required for bending machinery)

(6) Generate remaining corrugation

(7) Variation: panel with 3 corrugations.

(8) Variation: panel with 4 corrugations.

Information on the angle of the sun was fed into an algorithm that generated the bend profiles of the ceiling panels. The basic principles of the algorithm involved using the corrugations to block direct sunlight, while letting indirect light bounce and filter down to the space below. The top row of diagrams establishes the rules for the corrugation algorithm, and the bottom row plays out the rules over a single panel.

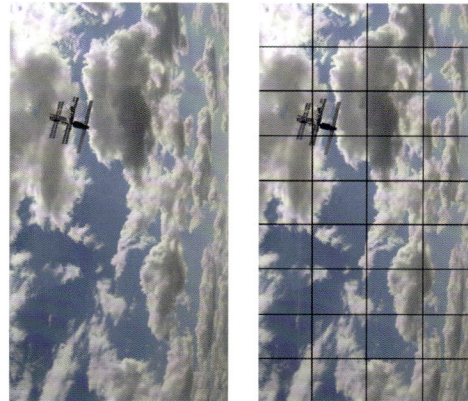

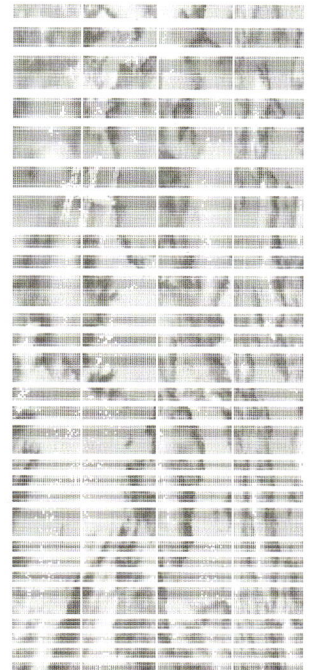

opposite top: The perforation pattern design involved calibrating the scripting process to respond to the pre-existing conditions in the ceiling, as well as various forms of infrastructure that would be integrated into it. Rules were developed to establish 'buffer zones' adjacent to light fixtures, sprinkler heads, edges of panels and the break/bend lines of the panels. Through a digital scripting process, the pattern generated from the acoustic analysis was modified accordingly to accommodate these rules.

opposite centre: The pattern for the west wall of the social hub was developed by filtering a photograph of the view outside the west wall of the Journalism School (a view across Broadway) through a digital process that converted it into perforations to be cut out of steel panels. The intention was not only to project the image of the outside on to the inside of the wall, but to also allow for different perceptions and readings of the wall depending on the viewer's proximity. The pattern was calibrated to 'snap' into focus at a specific distance of 12 metres (40 feet) – the point at which one walks into the social hub.

opposite bottom: The pixilation scripting process uses an alphabet of six discrete characters, each corresponding to a specific range of tonal values within the black-to-white spectrum. Through a simple algorithm, the script converts the image into a perforation pattern by replacing the raster information with characters according to tonal value. The characters consist of a perforation gradient from zero (0) to one (1) – the most basic forms of digital information.

above left: The ceiling for the new café addition is a sunscreen that hangs below a glass roof, and was designed and engineered to reduce heat loads. Two patterning techniques – corrugation and perforation – were used on the panels to develop the most efficient means of solar shading. Both techniques were developed in tandem through a solar analysis and scripting process to optimise solar shading and reduction of solar heat gain inside the new building, while also achieving the qualitative effect of being under a canopy of trees.

above: Once the corrugation pattern was determined, the resultant geometry was fed back into the energy analysis software. Each face of the corrugated surface was then assigned a maximum allowable percentage of perforation that would satisfy the solar heat reduction requirement. The pattern itself was derived from an image of the sky as if looking straight up from the café through the roof. The size and geometry of the perforations were determined by balancing the need for a resolution that would allow the image to be legible with the cost of laser-cutting the holes in the panels. **ᗄ**

Contributors

Craig Brimley is NBBJ's BIM integrator/designer, and works with the design industry using 3-D design tools to enhance design, constructability and system efficiency to increase value for the client. **Jorge Gomez** is a senior associate at NBBJ, and delivery leader for the New York studio, where he directs the application of BIM platforms on all of the firm's projects and works with NBBJ's Delivery Leaders Collegial Group, which is shaping innovative approaches focused on integrated project delivery.

Martin Doscher is the IT/CAD Manager at Morphosis, responsible for the day-to-day management of Morphosis IT infrastructure and support, as well as for overseeing CAD production for the practice and its consultants. His main focus is the integration of virtual building models into design and construction processes. He is leading the design teams for the Wayne Morse US Courthouse, San Francisco Federal Building, the University of Cincinnati Recreation Center and the New Academic Building for the Cooper Union in their efforts to integrate the architectural, structural and MEP 3-D models with the contractor's shop drawings. He has a Bachelor of Science degree with a major in architecture from Georgia Institute of Technology, and a Master of Architecture degree from SCI-Arc. He has been a guest lecturer and has conducted numerous workshops on BIM and parametric design at conferences including SmartGeometry, ACADIA and ZweigWhite's AEC/IT Strategies, and at universities including Auburn, USC and Yale.

Chuck Eastman is Professor of Architecture and Computing at Georgia Institute of Technology where he directs the AEC Integration Lab. His research assistants are **Jin-kook Lee**, **Hugo Sheward** and **Jaemin Lee**, who are all PhD students in design computing. **Yeon-suk Jeong** is a postdoctoral fellow in the College of Architecture at Georgia Tech. **Paola Sanguinetti** is an associate professor at the School of Architecture at the University of Kansas, and is currently studying for a PhD, and **Sherif Abdelmohsen** is Assistant Lecturer at Ain Shams University, Egypt. www.coa.gatech.edu/phd

Janet Echelman reshapes urban airspace with monumental public sculptures made of diaphanous, flexible surfaces that move and change shape through time. Major art commissions include the Richmond Olympic Oval, a venue for the Vancouver 2010 Winter Olympics, and the Phoenix Civic Space, a new signature icon that spans two downtown city blocks. Her recently completed *She Changes*, a 49-metre (160-foot) tall waterfront wind sculpture in Porto, Portugal, received the IFAI International Achievement Award and the Public Art Network's Year in Review Award. Her team won the Hoboken September 11th Memorial competition, which will result in the construction of a new free-standing island in the Hudson River. She was the only artist among the 2007–08 Loeb Fellows at Harvard University, and among the Henry Crown Fellows at the Aspen Institute, 2004–07. Recipient of a Fulbright

Lecturehip, Rotary International Fellowship, and grants from the New York Foundation for the Arts, Pollock-Krasner Foundation and Japan Foundation, she currently serves on the board of the Fulbright Association and on the jury for the Aspen Institute Energy and Environment Awards.

Richard Garber is an assistant professor at the New Jersey School of Architecture (NJSOA), where he teaches design studios and directs the school's FABLAB. His work involves the use of computer simulation and computer numerically controlled (CNC) hardware in the generation of innovative design, construction and assembly solutions. In 2007 his practice, GRO Architects, won the Lower Manhattan Cultural Council's re:Construction Competition. The resulting work, Best Pedestrian Route, was fabricated at NJSOA's FABLAB and was installed at the corner of Broadway and John Street in Lower Manhattan. In 2008 GRO won an AIA Merit Award and a NY Designs Award from the Architectural League of New York for these efforts. He was also the 'Emerging Architect' Visiting Assistant Professor at Barnard College in 2007, with Nicole Robertson. He was previously a project manager at SHoP Architects, where he worked on the firm's 2000 winning PS_1 entry, Dunescape, and at Greg Lynn FORM where he worked on the Presbyterian Church of New York. His writing and design work has been published in the *New York Times*, the *Star Ledger*, *The Architect's Newspaper*, *Azure*, *Art News*, *Metropolis* and *Architectural Record*. He holds architecture degrees from Rensselaer Polytechnic Institute and Columbia University. www.groarc.com

Urs Gauchat is Dean of the New Jersey School of Architecture at New Jersey Institute of Technology (NJIT) and a practising architect. The New Jersey School of Architecture is one of the largest in the US, and under Gauchat has been transformed into an internationally recognised leader in the area of CAD (computer aided design) and community development. He is particularly interested in creating a bridge between the considerable resources of universities and the needs of communities, and previously taught at the Harvard GSD and was president of the Boston Architectural Center. He currently serves as an adviser to academia, governments, communities and industry. architecture.njit.edu

Douglas Gauthier is the founding partner of Gauthier Architects, and was also a founding partner of System Architects. Prior to that, he collaborated with Frank Barkow and Regine Leibinger on two competition-winning schools in Berlin. He was previously a project architect on Le Fresnoy, Filmschool/Mediateque for Bernard Tschumi Architects and on the San Jose Repertory Theater for Holt Hinshaw Jones:Architecture. He has degrees from Columbia University GSAPP and the University of Notre Dame, and awards include a Fulbright Scholarship to the Slovak Republic, the Berlin Architekturpreis, a

Graham Foundation grant, and a MacDowell Colony Fellowship. He has taught widely in both the US and Europe. With Jeremy Edmiston he built a BURST* prototype in Australia, which was later commissioned by the Museum of Modern Art. A version of the house, BURST*008, appeared at full scale in the recent MoMA exhibition 'Home Delivery: Fabricating the Modern Dwelling'. Current projects at Gauthier Arthitects incude a performance theatre in New York and a research and exhibition space in Berlin.

Gary Haney is a design partner at Skidmore, Owings & Merrill (SOM). His building designs increasingly integrate algorithmic frameworks that serve as the basis for exploring form in relation to its associated performance. This research-intensive, performance-driven approach challenges the physical, structural and programmatic parameters of a given project and has enabled him to design buildings of great scale and complexity. His recent work includes the Al Rajhi Bank Headquarters in Riyadh, Saudi Arabia; the Al Hamra Tower in Kuwait City; and the Al Sharq Tower in Dubai, which was recently honoured with a Progressive Architecture Award.

Keith Kaseman received a BSD in architecture from Arizona State University and a Master of Architecture from Columbia University's GSAPP. From 2001 to 2003, he worked as a designer/project manager at SHoP Architects in New York. In June 2003, Keith and his partner Julie Beckman launched Kaseman Beckman Advanced Strategies (KBAS) upon having their scheme selected as the winning proposal in the Pentagon Memorial Design Competition. In 2006 KBAS, based in Philadelphia, was selected for the Architecture League of New York's Young Architect Award. Keith is currently a visiting lecturer in the University of Pennsylvania's Department of Landscape Architecture and an adjunct associate professor of architecture at Columbia University's GSAPP. KBAS operates under the premise that, at its best, architecture stands as a declaration of collaborative intelligence and exerts a positive force in the world. In this light, KBAS develops advanced strategies to further contribute to the design of our cultural fabric and beyond. www.kbas-studio.com

Stephen Kieran and **James Timberlake** are partners at KieranTimberlake, an award-winning and internationally recognised architecture firm noted for its research, innovation and inventive design. In 2001, they were the inaugural recipients of the prestigious Benjamin Latrobe Fellowship for Architectural Design Research from the AIA College of Fellows. In 2008, KieranTimberlake received the Architecture Firm Award, the highest honour bestowed on a firm by the American Institute of Architects. Their latest book, *Loblolly House: Elements of a New Architecture* (Princeton Architectural Press, 2008), is a case study of a single building which inaugurates a new, more efficient way of constructing off site through the use of building information modelling and integrated component assemblies. Stephen Kieran received his bachelor's degree from Yale University, and his Master of Architecture from the University of Pennsylvania. He is also a recipient of the Rome Prize, American Academy in Rome (1980–81). James Timberlake received his bachelor's degree from the University of Detroit, and his Master of Architecture from the University of Pennsylvania. He is also a recipient of the Rome Prize (1982–83). Both Kieran and Timberlake have served as Eero Saarinen Distinguished Professor of Design at Yale University, held the Max Fisher Chair at the University of Michigan, and taught at Princeton University. They currently lead a graduate research studio at the University of Pennsylvania's School of Design, and are endowed professors in sustainability at the University of Washington College of Architecture and Urban Planning. Recent projects include the Sidwell Friends Middle School, Yale University Sculpture Building and Gallery, the West Campus Residential Initiative at Cornell University, a post-Katrina home for the Make It Right Foundation, and Cellophane House, an off-site fabricated dwelling for the Museum of Modern Art.

Greg Lynn is the founder of design practice Greg Lynn FORM based in Venice, California. He graduated from Miami University of Ohio with degrees in architecture and philosophy, earned his MArch from Princeton University, and received an honorary doctorate from the Academy of Fine Arts & Design in Bratislava. In 2002, he left his position as the Professor of Spatial Conception and Exploration at the Swiss Federal Institute of Technology, Zurich, and became an ordentlicher university professor at the University of Applied Arts in Vienna. He is a studio professor at UCLA and the Davenport Visiting Professor at Yale University. In 2001, *Time* magazine named him one of 100 of the most innovative people in the world for the 21st century, and in 2005 *Forbes* magazine named him one of the 10 most influential living architects. He has received awards from the AIA, Progressive Architecture and the American Academy of Arts & Letters. The National Building Museum in Washington DC recognised him for 'shap[ing] the public realm' through his Korean Presbyterian Church of New York, which is also listed by the NYC Landmarks Commission. His work is in the permanent collections of the CCA, SFMoMA and MoMA. He also represented the US in the American Pavilion of the 2000 Venice Architecture Biennale. He is the author of seven books and monographs, including *Animate Form* (Princeton Architectural Press, 1997), which is considered the seminal book on the use of digital technology in architecture, and *Greg Lynn FORM* (Rizzoli, 2008).

Marble Fairbanks is an architecture, design and research office founded in 1990 by **Scott Marble** and **Karen Fairbanks**. Over the past years, Marble Fairbanks has received many local, national and international design awards including an Art Commission of New York City Award for Excellence in Design, AIA awards, American Architecture Awards, a PA Award, an ID Award, and an ar+d Award for Emerging Architecture from *Architecture Review* magazine. In 2004 Scott Marble and Karen Fairbanks were the Charles and Ray Eames Lecturers at the University of Michigan, and the book *Marble Fairbanks:Bootstrapping*, was published on the occasion of that lecture. The work of Marble Fairbanks is published regularly in journals and books and has been exhibited in galleries and museums around the world including the Architectural Association in London, the Nara Prefectural Museum of Art in Japan and the Museum of Modern Art in New York where their drawings are part of the museum's permanent collection, Fairbanks is Chair of the architecture programme at Barnard and Columbia College in New York, and Marble teaches at the Graduate School of Architecture at Columbia University where he directs the Avery Digital Fabrication Research Lab.

Nat Oppenheimer is a principal with Robert Silman Associates, a structural engineering firm with offices in New York, Washington DC and Boston. He joined the firm shortly after graduation from Clarkson University in 1988. Aside from his work with Robert Silman Associates, he is a lecturer at Princeton University and has taught or been a visiting critic at the architecture schools at Columbia University, Parsons, Pratt and the University of Pennsylvania. www.silman.com

Cynthia Ottchen is a registered architect and published author whose work focuses on innovative digital design methodologies, especially new generative strategies. She is the 2008 Gerald Sheff Visiting Professor in Architecture at McGill University and has been Head of Internal Affairs at OMA in Rotterdam, directing research and innovation for the office and initiating its Parametrics Cell. Her postgraduate studies at the University of Cambridge focused on new modes of thought and their effect on architectural design. Prior to joining OMA, she developed an innovative multimedia project in conjunction with Sony/BMG that incorporates new digital processes in music, spoken word and film.

Hani Rashid and **Lise Anne Couture**, founders and principals of Asymptote Architecture, are leading architectural practitioners of their generation whose innovative work and academic contributions have received international recognition. Founded in 1989, Asymptote is currently working on a broad range of commissions at sites in the US, Europe and Asia, including two commercial office towers in Budapest and a commission to design the World Business Center Solomon Tower in Busan, South Korea. The design for the Solomon Tower consists of three separate, tapered towers soaring out of a single base and at 560 metres (1,837 feet) will be among the tallest buildings in Asia. Also in design are two contemporary art pavilions commissioned by the Guggenheim Foundation for the Cultural District of Saadiyat Island in Abu Dhabi, UAE. Other recent competition proposals from Asymptote include the winning entry for an iconic, 40-storey corporate headquarters in Tbilisi, Georgia, and a dramatic design for a new Guggenheim Museum in Guadalajara, Mexico. In 2004, Rashid and Couture were presented with the coveted Frederick Kiesler Prize for Architecture and the Arts in recognition of their contributions to the progress and merging of art and architecture.

Nicole Robertson holds a Bachelor of Arts from the School of Architecture at Princeton University and a Master of Architecture from the University of California Los Angeles. Prior to forming GRO, she worked as a project designer for the Embryological House at Greg Lynn FORM in Los Angeles, and later as a senior designer at Skidmore, Owings & Merrill (SOM) in New York where she worked on the Automated People Mover Station at Dulles Airport. She previously taught for three years as a full-time tenure-track assistant professor of architectural design and representation at the School of Architecture, Syracuse University, and is currently an adjunct assistance professor teaching design studios in the Department of Architecture at Barnard and Columbia Colleges, and Graduate Representation I in the Graduate School of Architecture, Planning and Preservation at Columbia University in New York.

Coren Sharples is a founding partner of SHoP Architects and SHoP Construction Services. She holds a Bachelor of Science from the University of Maryland's College of Business and Social Science (1987), and a Master of Architecture from Columbia University's Graduate School of Architecture, Planning and Preservation (1994) where she graduated with honours for excellence in design and was a recipient of the William Kinne Fellowship for postgraduate travel and research. SHoP's work has won numerous awards, has been published and exhibited globally, and is in the permanent collection of the Museum of Modern Art. Current projects include several multifamily residential buildings in and around New York City, including the LEED's certified Garden Street Lofts in Hoboken, New Jersey; the East River Waterfront Park in Lower Manhattan; and a new campus for Google in Mountain View, California.

Dennis Shelden is a founder of, and Chief Technology Officer for, Gehry Technologies. Prior to forming Gehry Technologies, he was Senior Associate and Director of Computing at Gehry Partners, where he was responsible for the management and strategic direction of the firm's computing programme, including project applications, process and software development initiatives, and research and development. He holds a Bachelor of Science in Art and Design, a Master of Science in Civil and Environmental Engineering, and a Doctor of Philosophy in Design and Computation from the Massachusetts Institute of Technology. He is currently Visiting Lecturer at MIT. www.gehrytechnologies.com

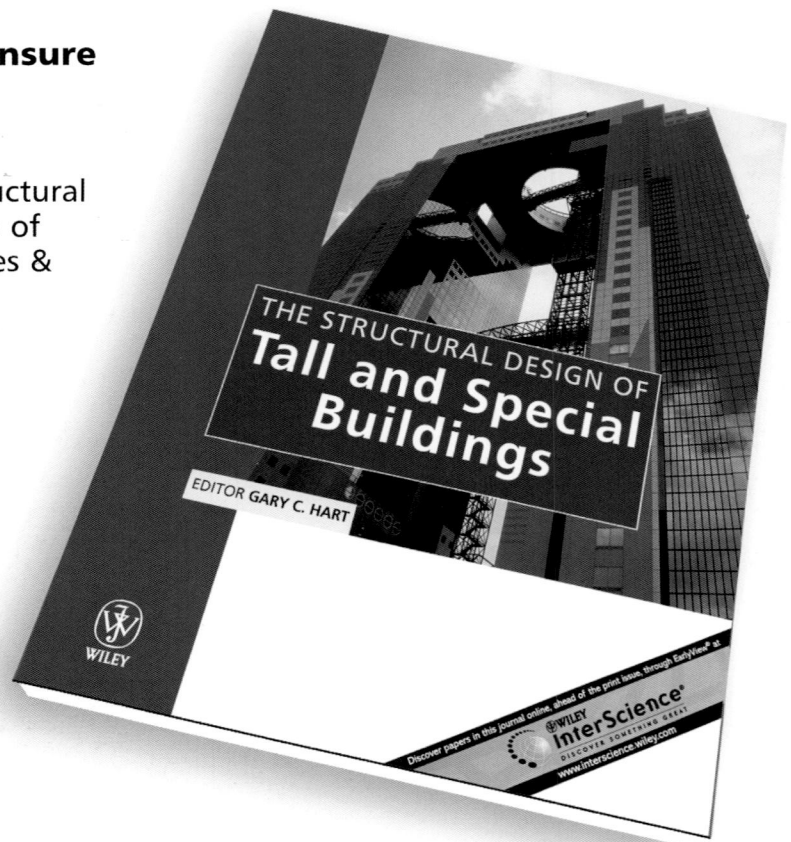

C O N T E N T S

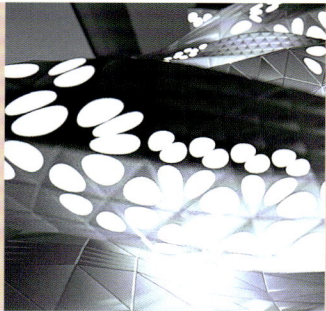

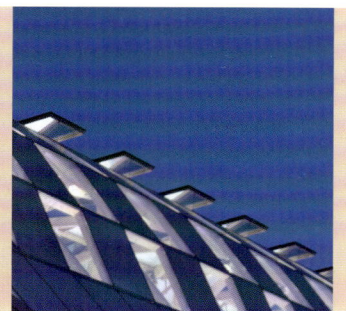

Carlos Zapata and Antonio Citterio's Cooper Square Hotel New York

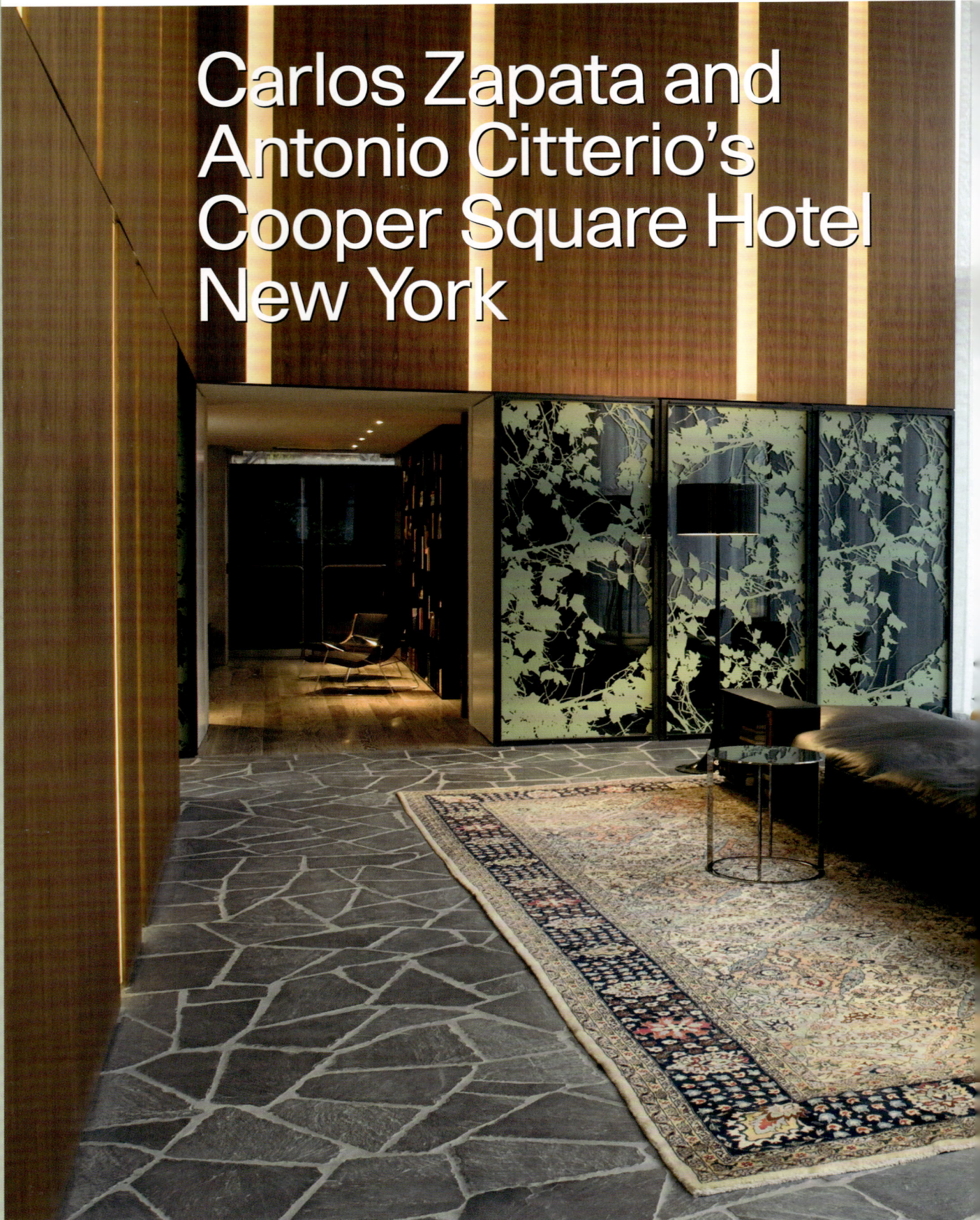

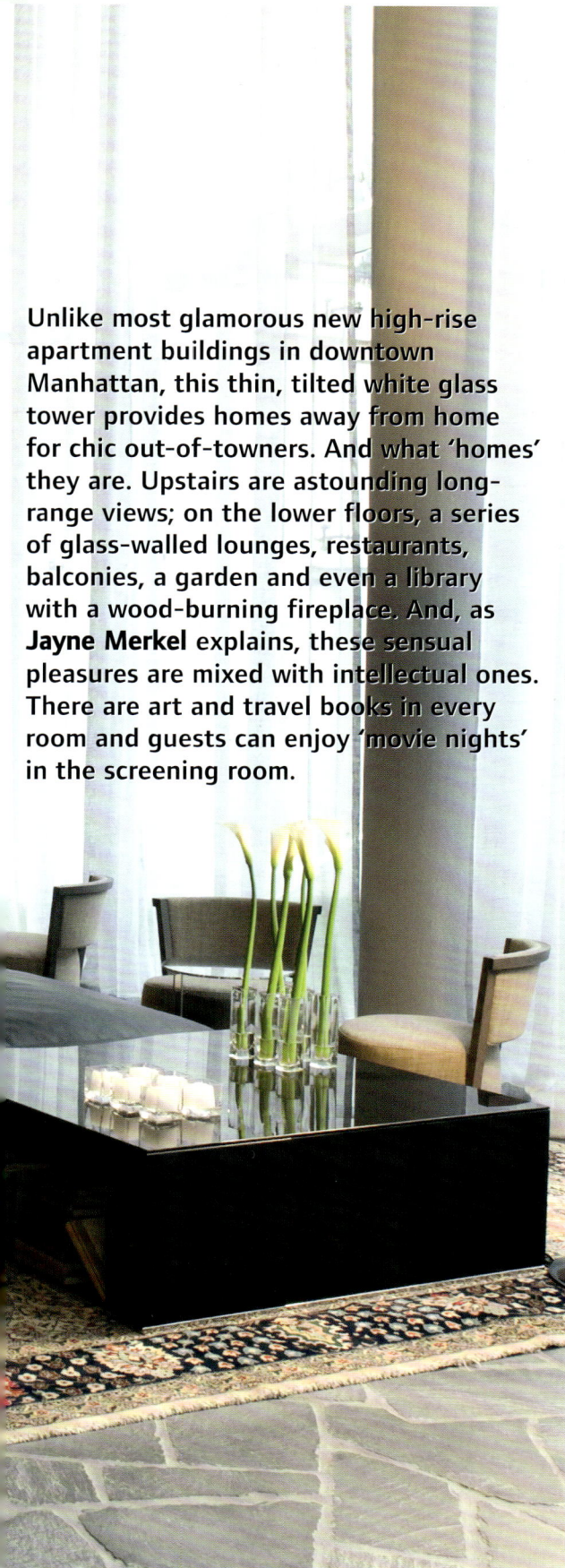

Unlike most glamorous new high-rise apartment buildings in downtown Manhattan, this thin, tilted white glass tower provides homes away from home for chic out-of-towners. And what 'homes' they are. Upstairs are astounding long-range views; on the lower floors, a series of glass-walled lounges, restaurants, balconies, a garden and even a library with a wood-burning fireplace. And, as Jayne Merkel explains, these sensual pleasures are mixed with intellectual ones. There are art and travel books in every room and guests can enjoy 'movie nights' in the screening room.

On a wide stretch of the once-tawdry Bowery, this glamorous new hotel has opened just south of the quaint 1859 brownstone edifice that houses the Cooper Union, an influential art, architecture and engineering school that has long been a training ground of the architectural avant-garde. John Hejduk, who served as dean for decades, created some of the city's most spectacular white modern interiors in the historic structure where Abraham Lincoln once spoke. Now Anthony Vidler is the dean, and an almost-completed angular new Cooper Union Administration Building by Thom Mayne of Morphosis is going up next to the new hotel.

Only a few years ago the neighbourhood was filled with tenement buildings, offbeat nightspots and wholesalers, but now SANAA's New Museum of Contemporary Art is only a few blocks away, Norman Foster has been hired to design a nine-storey building for the Sperone Westwater Gallery just north of it, and the increasingly fashionable SoHo and Greenwich Village are nearby. So this is a perfect location for a hotel targeted at an art, fashion, media and movie crowd. Its first party celebrated the debut of *The Duchess*, the Keira Knightly film based on Amanda Forman's celebrated biography of Georgiana, the 18th-century Duchess of Devonshire.

Carlos Zapata Studio's dramatic torqued 20-storey tower provides unobstructed views, not only of lower and midtown Manhattan, but of much of Brooklyn and Queens. You can see all the way to the airports if you squint. The views are the main event in the décor of the rooms, though the sleek furnishings were designed by Antonio Citterio and Partners of Milan. It is their first hotel design in America, and everything in the rooms had to be freestanding because of the building's unconventional shape. It is on the public rooms of the lower floors that they really get to strut their stuff, although subtlety rules here too.

There is nothing subtle about the entrance, however. A sleek glass canopy extends from the street through a geometric garden that screens the building's glass-walled lounges from the sidewalk. Visitors enter between a dramatically angular four-storey building on the left (with a tilted column in the corner and a restaurant on the ground floor) and the main hotel tower on the right, where 4-metre (14-foot) tall walnut doors lead to the lobby. Visitors can also go straight upstairs to another cocktail lounge with outdoor seating that hovers over the rear garden. Here they can gaze down into the screening room above the restaurant, up at the tower, or enjoy more layered grittier views of surrounding tenements which have a lively mix of tenants. Two poets living on the second floor of the one just south of the hotel refused to move, so the owners bought the building and built the hotel around it, placing the new hotel's library on its ground floor.

Carlos Zapata Studio and Antonio Citterio and Partners, Cooper Square Hotel, The Bowery (Third Avenue), East (Greenwich) Village, New York, 2008
Huge walnut doors lead from a walkway covered with a glass canopy into the lobby lounge. There is no reception desk here; clients are greeted subtly as if they are tenants. Double-paned glass walls stencilled with silhouettes of foliage provide a counterpoint to real foliage visible through the clear-glass west wall.

Carlos Zapata Studio and Antonio Citterio and Partners, Cooper Square Hotel, The Bowery (Third Avenue), East (Greenwich) Village, New York, 2008
The hotel's library is actually on the ground floor of a tenement from which the poet tenants could not be persuaded to move, so the hotel was built around it. All the books – mostly on art, architecture, design and travel – came from Housing Works, a nonprofit institution committed to ending AIDS and homelessness. The organisation runs a large bookstore café with donated goods and will replenish the shelves of the hotel as guests remove the books.

The gently twisted clear-glass and fritted tower of the Cooper Square Hotel tilts slightly as it rises, creating an intriguing presence on the skyline and providing larger rooms on higher floors where the views are unobstructed.

All guest rooms have tilted walls of glass with operable windows and dramatic views. Furnishings designed by Antonio Citterio and Partners are freestanding since room shapes are irregular. Every room has fresh flowers, a workspace with a bookshelf stacked with books and magazines, and a desk with an iPod docking station and a 40-inch flatscreen TV developed for the hotel with Sony. Though the hotel is pet friendly, the beds have 400-thread count Italian Sferra linens.

opposite: The most sensual rooms in the Cooper Square Hotel are the dark-walled bathrooms with freestanding tubs, glass-walled showers, Italian glass walls and ceilings and porcelain floor tiles. They are all stocked with lush linens and bath amenities created exclusively for the hotel by Red Flower. Some rooms even have views over the city.

The book-filled library though, like everything else in the Cooper Square Hotel, has an entire wall of glass. This one looks out on the rear garden on the east side. The hotel's main public space, the lobby off the outdoor covered entryway, has a glass west wall facing the street-front greenery. It has no reception desk. Hotel guests, like residents of New York's doorman buildings, will be recognised by the staff outside and subtly slipped keys in one of the lounges.

The lobby's east wall is panelled in walnut, with vertical recesses painted white and containing invisible (unless you stick your head in the narrow recess) light tubes to create light-filled voids. The floors are black Italian slate, laid in an irregular pattern. The north and south walls are double-paned glass stencilled with a leaf pattern in light-green paint, a motif that is repeated on the wallpaper of the upstairs corridors. These stencilled murals, though, are the closest the hotel comes to having art on its walls. All aesthetic effect derives from the architecture, views and gardens, and the books on shelves everywhere.

Every one of the 145 rooms, most of which are rather small, has a bookcase as well as a desk, a 40-inch flatscreen TV, and a modern armoire with hanging space, drawers, a fridge/mini-bar and a safe. But the *piéces de résistance* are the views through the double-paned operable windows, sheltered by sheers and opaque drapes, and the dazzling dark-walled bathrooms with freestanding tubs and separate glass-walled showers.

A 149-square-metre (1,600-square-foot) two-bedroom glass-walled suite on the 20th floor has a freestanding fireplace, two bedrooms and two-and-a-half bathrooms. It opens on to a wraparound terrace of the same size, with some of the most spectacular views in New York. It is here that the rationale for the enticing shape of the gently twisting tower becomes clear. Each floor offers a slightly different series of views. The building has a larger floor plate at the top, where the views are more spectacular. 'Like a face, it gets wider after the neck,' Carlos Zapata explains. The profile appears slimmer from the front (west) side.

White frits in a dotted pattern, with dots of various sizes, provide sunscreening and give the glass walls a sense of solidity so the outer walls can catch the light. In some areas, the dots are terracotta coloured, in deference to the surrounding brick building stock. Similarly, the colouration of the green cube near the entrance was inspired by the copper dome of a nearby 1867 Ukranian church which is visible from above. White-painted aluminium panels on the back of the hotel are imprinted with a dotted pattern similar to the fritting, so the inspiration doubles back on itself.

The Cooper Square Hotel currently stands out in this neighbourhood, though Gwathmey Siegel's serpentine apartment tower on Astor Place is only a few blocks north. But when the Morphosis building opens next door, it will be part of a cluster of dramatic 21st-century buildings integrated with the 19th-century ones that form most of the area's fabric, giving their sleek new brethren a certain rooted footing in the ground. Δ+

Pierre Thibault

At a time when little in architectural design seems like uncharted territory, **Brian Carter**, Professor and Dean of the School of Architecture and Planning at the University of Buffalo, The State University of New York, goes northeast to Quebec and encounters the work of architectural explorer Pierre Thibault. He describes how Thibault sets out to develop projects in some of the most 'remote, expansive and sparsely settled' areas of the region, responding afresh to the landscape and its indigenous culture.

Jardins d'hiver, Lac Turgeon, Charlevoix, Quebec, 2001
Designed and constructed by a team of architects, artists and students, the project sits within a mountainous region with an altitude of more than 800 metres (2,625 feet) and which experiences frequent gale-force winds. Temperatures often drop to −40°C in this part of Quebec and the lakes often remain frozen throughout the year. This is the southernmost taiga in the world and the gardens, which are defined by black spruce forests, are focused around a trail that connects seven lakes. The project was part of a programme to enhance the trail, which is open to the public throughout the year and extends over several kilometres through the mountains.

The work of the French-Canadian architect Pierre Thibault is rooted in the territory where he practises. Born in Montreal and a graduate of the architecture programme at Laval University, he first established an office in Quebec City in 1988 and, in the last 20 years, has gone on to develop a unique portfolio of work that reflects the particular characteristics of that place.

The province of Quebec defines the northeastern corner of North America. Straddling the St Lawrence River and with a western boundary that partially defines Hudson Bay, the province extends to Newfoundland in the east. In the south it not only defines the outskirts of Canada's federal capital of Ottawa, but forms a part of the country's international border with the US, while to the north it extends almost to the Arctic Circle. More than twice the size of France and focused around two major cities – Montreal and Quebec City – it is made up of predominantly wild natural landscapes and defined by radical seasonal changes of extreme weather. First settled by indigenous tribes, the vast province was also a home to both the British and the French. As a consequence it is steeped in a long and particularly unique history defined by aboriginal and foreign languages and a dense mix of different cultural traditions.

After establishing a cluster of fortified buildings at Cap Diamant as a tentative outpost of New France in North America, in 1608 the French explorer Samuel de Champlain embarked on an arduous journey to transform this particular territory into a colony. And, almost 50 years later, when a group of French mystics sought to establish a missionary city in the wilderness, they chose a site in Quebec that was subsequently to become a foundation for the city of Montreal.

Modern architecture in French Canada has been referenced by its 'apparent absence'[1]– a characteristic that is perhaps a consequence of the location of Quebec, its vastness and those cultural forces which, over time, have tended to combine to shield it from view. And while it is perhaps understandable that architecture might appear to be absent in such a remote, expansive and sparsely settled part of the world characterised by vast, icy landscapes, this is also a place where art and architecture have more recently been conspicuously cultivated and focused in an attempt to define a distinctly different culture.

Pierre Trudeau likened Canada's relationship with its large and powerful southern neighbour to 'a mouse in the shadow of an elephant'.[2] While the architectural community in Canada is small compared to that south of the border, architects there have been working hard to resist the homogeneity that characterises much of the built environment of the US by defining an architecture in the north that is particular to the region.

Thibault's work is notable in this context, as he has focused his attention not only on the design and construction of buildings, but on studies of the land. To advance these studies, he has undertaken a series of projects over the last few years that have been designed to respond to the remoteness and vast scale of Quebec's natural landscapes. Developed as an initiative that not only seeks to build an appreciation of the impact of land and weather, this is also one that embraces the education of architects and artists.

For each of the projects he sets out with a small team to develop responses to vast tracts of land. Often remote and far from any obvious sign of civilisation, the sites become the inspiration for design. Situated between Land Art, sculpture and architecture, they invariably prompt the construction of an installation that is designed 'by the site'. The sites, which frequently include both land and water, have in turn prompted initiatives that suggest ways of both measuring and occupying space. Using palettes of materials often determined by what is available – ice, fire, water, wooden sticks and light sheet materials – Thibault and his teams have designed and constructed primitive shelters and created sources of light and grids of markers that bring to mind early surveys and measures of the land. They also reference the idea of humans as shapers of the land, and work by artists such as

Beaver Lake Refuge, Grandes-Piles, Comple Laviolette, Quebec, Canada, 2000

The house is located in a dense forest of maple and pine trees and has been carefully integrated into the forested landscape. Constructed on an expansive deck of ash wood, which creates a new datum that hovers above the sloping site, it is clad in eastern white cedar and consists of a series of rooms grouped under a single flat roof supported on treated grey pine logs which are clustered to define spaces on the external porch. The living room has been planned within a double-height space with views out to the forest, and the main fireplace, which is constructed of limestone from Saint-Marc-des-Carrières, creates an internal focus.

Walter De Maria, Alice Aycock, Andrea Zittel and Mary Miss, as well as recalling those initial tentative moments of discovery that preceded inhabitation and the settlement of these strange new worlds.

Thibault stresses the importance of directly experiencing the place. He enjoys slowness and is suspicious of much of the speed that now characterises our lives and shapes current practices of architecture. Consequently, his projects are organised around carefully planned expeditions to the different sites and offer a view of the architect not merely as an artist and a builder, but as an explorer. The places studied are remote and Thibault and his team often have to trek to the sites and set up camp on arrival. This is demanding and requires careful planning, energy and the thoughtful consideration of the basic aspects of life that are often taken for granted – clothing, shelter, materials, food and supplies. The design team is responsible not only for specifying materials for the projects, but also for packing and carrying them to the site. Consequently the weight, shape and size of materials, along with the choice of tools for on-site fabrication in extreme conditions, suddenly become critical concerns.

Two of Thibault's projects – Jardins d'été (2005) and Jardins d'hiver (2001) – were situated in the Charlevoix region of Quebec. Designed to celebrate the vast natural landscape and the different seasons in the area, they defined the frozen surface of the land and a large lake with an orthogonal grid of markers and simple shelters. Lit by candles, which flickered in the wind, the new markers mapped out the extent of the frozen territory by developing a primitive survey-like tool constructed on site by the designers. Such interventions created a stark contrast with the flowing lines of nature and recalled the ordering systems devised by the first settlers. The two projects were also influenced by studies of the techniques and skills developed by the indigenous populations that have occupied this territory for centuries.

In addition to these studies of the natural landscape, Thibault and his team have also designed a number of buildings that celebrate the tenacity and history of the indigenous peoples. These include the new Anenakis Museum in the first nation village of Odanak on the South Shore of the St Lawrence River, the Dame de Coeur Theatre – a large skeletal enclosure of wood and steel planned to accommodate giant-sized puppet productions – in Upton, Quebec, and the Centre National des Naufrages du Saint-Laurent, a facility whose design recalls the wooden-fenced enclosures that defined indigenous settlements.

The settlement of the land, and the need to respond to particularly remote settings, have also shaped many of the houses Thibault has designed. A recent project, Les Abouts, was planned as a country house on a site that was bounded by a river and dominated by a dense pine forest. By contrast, the earlier Beaver Lake Refuge (2000) was developed as a refuge at the centre of a 2.6-square-kilometre (1-square-mile) woodland site close to a lake at Grandes-Piles in Comple

Cistercian Abbey of Saint Jean-de-Matha, Quebec, 2009
Currently under construction, Pierre Thibault's collaborative design (with Claude Demers and André Potvin) for the new abbey seeks to embody the harmony and inner calm that is at the centre of the daily routines of the monks who will live there. Contemplation, hospitality and prayer all hold privileged places for the monks, and in this setting time is also marked by the different seasons. Within this context the design of the abbey addresses the complementary imperatives of introversion and looking outwards, and the plan has been developed so as to create a series of linked spaces that are open to the sky and integrate landscape and buildings. The circulation has been organised to prompt movement around these open spaces, which have been planned on different levels and framed by a series of open structures offering views out and into the forests beyond.

Laviolette, Quebec. This was where the first builders were beavers, and Thibault speaks of the loose collections of sticks that made up their lodges on the lake and the landscapes of the forest as inspiration for the design of this new single-storey house. Constructed of wood, the house consists of a number of enclosed spaces grouped under an expansive roof and supported by groups of roughly hewn tree trunks.

This permanent residence, designed to accommodate a family and friends, is used throughout the year. Planned on an expansive wooden platform that appears to hover above the uneven ground, it provides a total of 372 square metres (4,000 square feet) of enclosed space together with a collection of outdoor spaces marked by tree-like clusters of wooden columns in ways that place the house ambiguously within the surrounding forest. One of a series of houses that have been designed by Thibault for a range of different sites in Quebec, this is a building without nostalgia for the land, yet a design that has been clearly inspired by the wilderness and the nature of its spectacular setting.

Thibault's work has also focused on the creation of places that recognise both the significance of the native people and the cultural history of French Canada. In this context the design of a new community for a group of Cistercian monks that is currently under construction in Quebec is arguably his most significant project. Planned to house a working community, the Cistercian Abbey of Saint Jean-de-Matha (2009) recalls the first settlements, like Cap Diamant. Remote from the city it has been planned to reference the expansive natural landscape that surrounds the new settlement. Contemplative outdoor spaces form the heart of the scheme and are defined by a series of austere yet elegant cells for the monks together with gathering spaces for the community. The project is being built of materials that also reference the immediate surroundings, with wood and stone combining to create thick walls that provide a response to the extremes of climate and at the same time provide interstitial spaces that can be readily used by this new religious community in the wilderness.

When asked if his interests in photography predated his interest in trekking, the renowned French-Canadian photographer Robert Bordeau spoke of how 'trekking – I call it wandering – came first. … I wanted to find out where I was in relationship to what was out there, so I wandered in order to find these things out.'[3] The work of Pierre Thibault clearly comes from a similar interest. By wandering, he has been able to establish views and accumulate experiences of the land that provide a better understanding of what is out there. Thibault has an obvious fascination with the natural landscape and a similar commitment to direct experience – of being there and of seeking out that heightened sense of awareness that comes from exploring a place. It is an unusual approach to architecture and design in a society that is increasingly shaped by speed, the anonymity of digital technology and the instant gratification promoted by the attention of the media. Yet it is one that clearly defines his artistic and architectural interventions and provides vital inspiration. It is also an approach that reveals the importance of rigorous studies of both physical and cultural territories as a basis from which to create a new modern architecture in French Canada. ∆+

Brian Carter is Professor of Architecture and Dean of the School of Architecture and Planning at the University at Buffalo, The State University of New York. Prior to taking up an academic appointment in North America he worked with Arup in London and has designed many award-winning buildings. The curator of several exhibitions of architecture and design in Europe and North America, he is also the author of numerous articles and books that have been published internationally. He has written extensively on modern architecture and is the editor of a series of books about architecture in Canada. He was also part of the design/research team for a new manufacturing facility for Ferrari in Italy.

Notes
1. Melvin Charney, 'Quebec's Modern Architecture', in Geoffrey Simmins (ed), *Documents in Canadian Architecture*, Broadview Press (Peterborough, ON), 1992, p 268.
2. *Time* magazine, 16 December 1974.
3. Robert Bordeau, *Industrial Sites: An Interview with Robert Enright*, Jane Corkin Gallery (Toronto), 1998, pp 15–16.

Herbert Art Gallery and Museum, Coventry

Pringle Richards Sharratt (PRS), Herbert Art Gallery and Museum, Coventry, 2008
The S-shape of the roof system can be clearly seen here. On the left is the exhibition block, clad in red Hollington sandstone.

Coventry's Herbert Art Gallery and Museum is a curious institution. It contains a little of everything: artworks, archaeological artefacts, bits of machinery, a history of Lady Godiva, the city archives, a missile, a medieval loom and many other curios. The building itself is a little odd too. When completed in 1960 it turned its back on the city's ruined cathedral and Sir Basil Spence's replacement, finished two years later. The idea of forming one corner of a new city square was not part of the building's role, and instead the principal entrance faced a new and rather busy road.

What Pringle Richards Sharratt (PRS) has done is turn the building around, in every sense of the phrase. The entire project was driven by a response to the medieval grid of prewar Coventry and views across to the cathedral; the original entrance has been closed off and access points located in quieter, pedestrian zones. It is also a more polite building – a little quirky, but something with personality rather than the monolithic (and, in parts, Brutalist) original. Designed by Albert Herbert, a relative of benefactor Sir Alfred Herbert, a good deal of the original building still survives, and earlier improvements by Haworth Tompkins helped steer the Herbert towards rehabilitation. PRS has gone much, much further and almost entirely reinvented it. It has been a long process. The practice was appointed in November 2002, but changes of mind and funding problems (partly due to political changes at the city council) caused one delay after another; PRS' first scheme, granted planning consent in August 2004, was scrapped and a replacement design was not approved until almost a year and a half later.

What makes this £11.5 million project so interesting is that it has been driven by the architectural programme rather than by the design language. 'We don't want to be dogmatic about these things,' says partner John Pringle. Painfully aware that Coventry was once a marvel of Medievalism, the architects have worked hard to respond to archaeological traces. The grain of a strip of town houses nearby, demolished in the 1950s, has been picked up by slabs of Corten steel that function as sculptures in a 'peace garden'. Moreover, these floor plans have been projected into the new Herbert building, giving much of the newbuild a distinctive twist, a certain obliqueness in plan that can only have given the engineers a headache.

The building runs north–south, but is doing everything it can to veer counterclockwise; the lateral elements have torn themselves free of an east–west straitjacket and now lie along an axis that is more east–northeast, aligning themselves with their medieval forebears. What this does is to also focus views slightly askance. Immediately to the north of the Herbert is a rather ugly student accommodation block, but the end wall of the new arcade is set (of course) along a diagonal, and therefore frames views of the cathedral. The Herbert tries to reverse the recent architectural history of Coventry; rather than begin at Year Zero, the building has a sense of the ground that it occupies. It therefore becomes something of a hybrid.

Structurally, too, it is a hybrid. Six years ago PRS envisaged a vaulted arcade covering the space between a pair of parallel, two-storey wings – one new, the other dating from 1964. The older wing came to be demolished and replaced by a single-storey building, so the canopy took on something of an S-shape, negotiating the space between

**Pringle Richards Sharratt (PRS), Herbert Art
Gallery and Museum, Coventry, 2008**

above: Ground-floor plan, showing context.
The original building runs east–west towards
the bottom of the drawing; PRS' newbuild
runs north–south. The skewed plan, which
takes its cue from the position of medieval
houses that once stood near the site, can be
clearly seen.

left: The Herbert Art Gallery and Museum
glimpsed from within Basil Spence's cathedral.

opposite: The similarities of the vaulted arcade
with PRS' Sheffield Winter Garden project are
striking, although the structural programme is
very different. Both projects provide semi-
public, enclosed routes from one part of the
city to another.

Views out of the history centre through to the cloister, peace garden and ruined cathedral. Corten-steel slabs mark where the walls of medieval houses stood until the 1950s.

The 'cloister' running alongside the single-storey history centre. The spruce of this structure contrasts with the precast concrete of the exhibition spaces.

buildings of different heights. At the lower end the roof is pure post and beam, but then it sweeps up and assumes something of a gridshell structure, the threads of which are aligned with both the cathedral and those lost medieval houses. There are other contrasts. The arcade and single-storey history centre use spruce as the structural material, but the two-storey exhibition block is concrete, exposed for its thermal mass and cooling properties. Some parts of the building are naturally ventilated; others are mechanically air-conditioned. There is an awful lot going on here.

So what stops this building becoming a mishmash of style, technique and response? It can only be the care with which this experienced firm of architects has approached the project. A scheme which blends subtlety and complexity like this can be achieved only by obsessing over the detail and keeping a firm grasp on the concept. PRS also has a strong sense of what is, and what is not, appropriate. Some of the fixtures of the original building (like the distinctive metal-framed windows and outsized architraves) have been left in place, while a handful of spaces have been entirely refashioned.

There is also a layering to the way the entire ensemble is perceived and experienced. On the one hand, the practice does not want the building to be considered clever or sophisticated – just a natural and comfortable place to be. 'The idea is that you don't feel intimidated. It's not a temple to high culture,' says Pringle. But on the other hand, there are ever-increasing levels of refinement which are there to be appreciated if only one stops to look for them: the plan as parallelogram rather than a rectangle, the ways in which the building is assembled from fragments (like Coventry itself), the reoriented views, the markings in the landscape which trace the footprints of disappeared buildings (and the granite line which plots the presence of a very real medieval vault below ground, now accessed through a corridor of the utmost austerity).

The Herbert is now a building that spills out into the landscape. A cloister runs beneath an overhanging roof containing no gutter (creating, when its rains, a waterfall which drops directly into an inconspicuous drain). The peace garden has been intelligently thought out, and the Corten sheets have been incised with the names of people who once lived in the houses that stood here. Amphitheatre steps deal with the changing topography while encouraging people to stop and sit. Gripes? Just one. The way that Event Communications has delivered the interpretation and curation of gallery objects is very often intrusive, heavy-handed and patronising – interesting objects can be lost among all the sound, colour and interactivity. But this is not the fault of PRS, which has endeavoured to create both a building and a civic space. The result is a vaguely eccentric and eminently likeable building which makes more and more sense the longer one looks at it. Δ+

David Littlefield is an architectural writer. He has written and edited a number of books, including *Architectural Voices: Listening to Old Buildings*, published by John Wiley & Sons (October 2007). He was also the curator of the exhibition 'Unseen Hands: 100 Years of Structural Engineering', which ran at the Victoria & Albert Museum in 2008. He has taught at Chelsea College of Art & Design and the University of Bath.

Computational Spring Systems
Open Design Processes for Complex Structural Systems

Sean Ahlquist and Moritz Fleischmann are both former graduates of the Emergent Technologies and Design Programme at the Architectural Association (AA) in London, who have now established their own innovative research and design practice, morse. Here they describe their work with computational spring algorithms that has enabled iterative investigations of tension active systems and structures.

In most design processes the generation of form can be distinct from the application of a structural strategy. Working with tensile structures, the link between material arrangement and the movement of structural force is intimately intertwined. This is a compelling proposition from an architectural standpoint. Shape and structure are formed simultaneously, but the process of developing such structures is highly technical and exhaustive if carried out through either analogue modelling and analysis techniques or in specialised engineering software. The breadth of knowledge necessary to utilise these methods and their highly prescriptive nature are impediments to a design process investigating variation in form of this structural type and its potential for nonstandard spatial and environmental conditions. The development of computational spring algorithms has provided an avenue for efficient studies of tension active structures. Accomplished through a programming language such as Processing (processing.org), the design environment is also accessible and variable. Tracking through examples of spring systems generating hybrid tension/compression structures, a design process can be formed with elemental means of geometry and programming, making possible iterative investigations of complex tension-active systems.

Element

A computational spring describes force acting between two particles of a particle system. A particle is the primary element of the system carrying a value for mass. The determination of position occurs once a force, in this case coming from a spring, acts upon it. The position of a particle can be calculated through the simple equation of $f = mass*acceleration$, where f is the vector sum of all forces.[1] A spring system follows Hooke's Law where force is related to the amount of displacement between two particles connected by a spring. Building a network of springs is a matter of associating an array of particles. This will signify the type of network topology that is formed. This topology is a user-defined condition and is nonpositional. The geometric form is determined algorithmically, through the method of dynamic relaxation. Vector forces oscillate between springs resolving the system to a force equilibrium state. This process converts the network topology into positional geometric form.

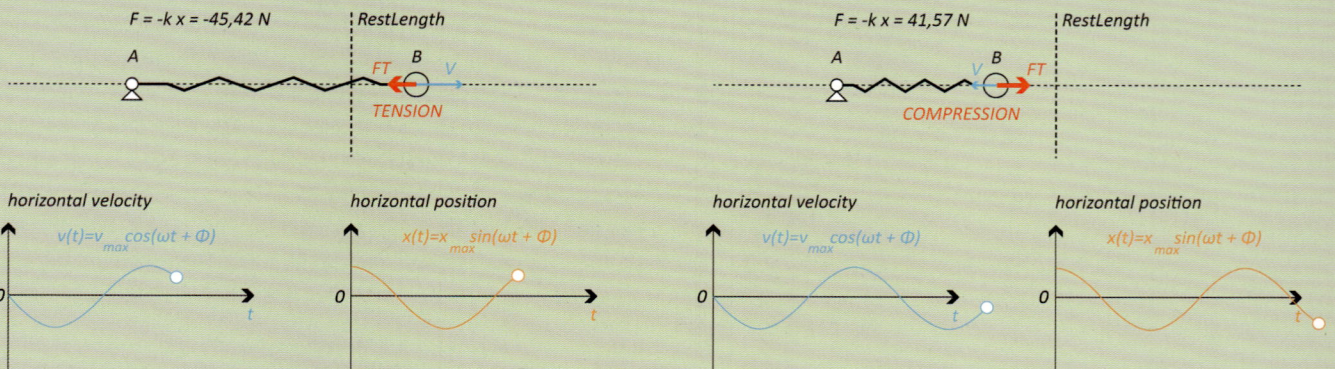

$F = -k\,x = -45{,}42\ N$ — RestLength — TENSION — $v(t)=v_{max}\cos(\omega t + \Phi)$ — horizontal velocity — horizontal position — $x(t)=x_{max}\sin(\omega t + \Phi)$

$F = -k\,x = 41{,}57\ N$ — RestLength — COMPRESSION — $v(t)=v_{max}\cos(\omega t + \Phi)$ — horizontal velocity — horizontal position — $x(t)=x_{max}\sin(\omega t + \Phi)$

morse, Diagram of Hooke's Law, 2008
A spring-based particle system. The spring emits force between the two particles based on Hooke's Law.
Where particle A is fixed, the position of particle B is determined by the oscillation of the spring force.

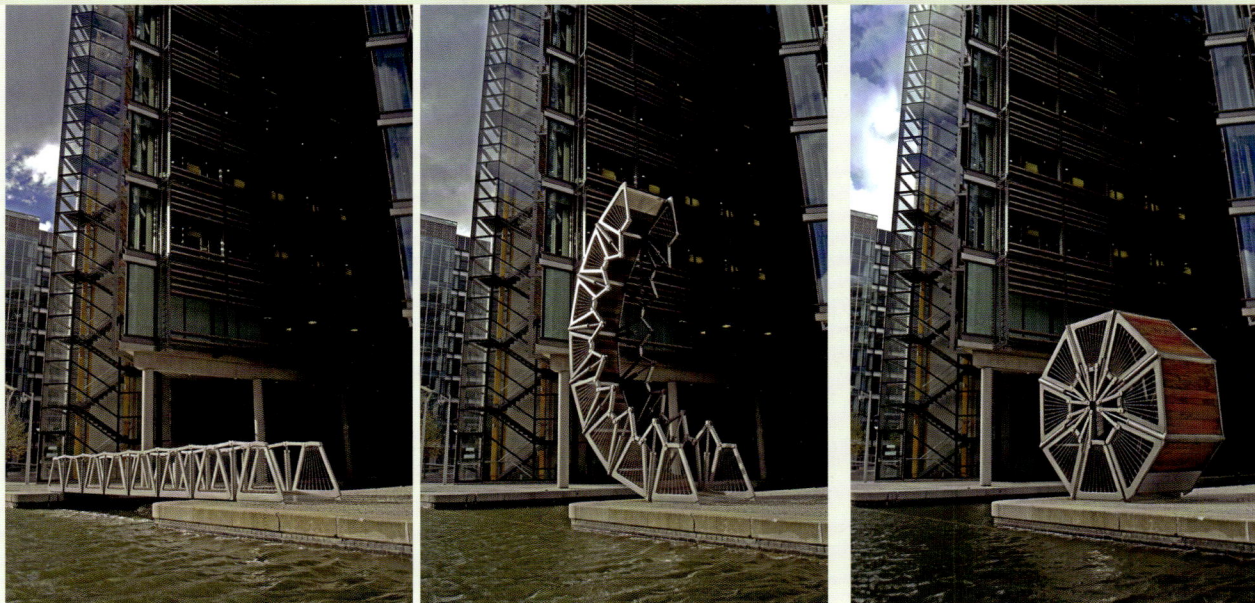

Heatherwick Studio, Rolling Bridge, Paddington Basin, London, 2004
The kinetic steel-framed bridge transforms through the hydraulic
expansion and contraction of various struts in the truss-like structure.

The association of springs – the network topology – is
critical while also powerful in the creation of a tension-
active system. This is due to the ability of a spring to be
realised as a tension or compression member. A spring
emits force when it has been displaced from its rest
length.[2] Rest length is, generally speaking, a constant,
established prior to the engagement of the forces from the
overall network of springs. When the springs interact, a
spring whose actual length is more than the rest length
will exhibit a pulling force (tension). A spring whose
length is less than the rest length will act in compression.

Mapping

Moving through the stages of development, the challenge
is how to track and control variation from the initial setup
through to the final form. Abstracting processes described
in evolutionary developmental biology provide the logic for
proceeding through the steps of geometric form
development. In defining the base geometric unit,
extensibility is primary. Embryological development works
with several primary materials to realise multiple
structural types (both tension and compression systems).
Instead of prescribing multiple context-independent
components, a single unit should be able to develop into
various material types.

The fate map, a representation technique for tracking
cell development, is a helpful concept in managing form
specification in complex systems. The control of
articulation can only be gained when the initial setup can

be connected to the final form. In the spring-based system, the
network arrangement is akin to the fate map, the particle being
equivalent to the cell. A logical ordering system for the
components of the map allows for easy determination of where
particular articulations in form occur and in what contexts.

Connecting the dynamic nature of spring systems and the
effects of mapping particular associations of springs is expressed
in a model that describes the Rolling Bridge by Heatherwick
Studio. By configuring the truss network with springs, the
simulation, written in Processing, is able describe tension and
compression forces, and the switching in between. The experiment
seeks to understand the physical behaviour of the constructed
bridge through the conceptual nature of the computationally based
spring system. This offers interesting possibilities in describing
valid structures through a computational component that has
specific structural properties but not absolute material properties.

Spring systems become effective design experimentation tools
when they are computationally efficient and openly accessible.
Visually, responsiveness is necessary to understand the
relationship between the outcomes emerging from the process and
the input parameters. Accessibility is necessary from a
programming standpoint, so that the algorithms necessary to run
the system can be developed and controlled by the designers
themselves. A design system developed by Soda shows how a
complex structure can be understood and refined visually through
interactive engagement and immediate feedback. Sodaconstructor
is a Web-based construction kit for spring systems, developed
through an interest in dynamic systems where behaviour could
arise through the feedback between the user and the

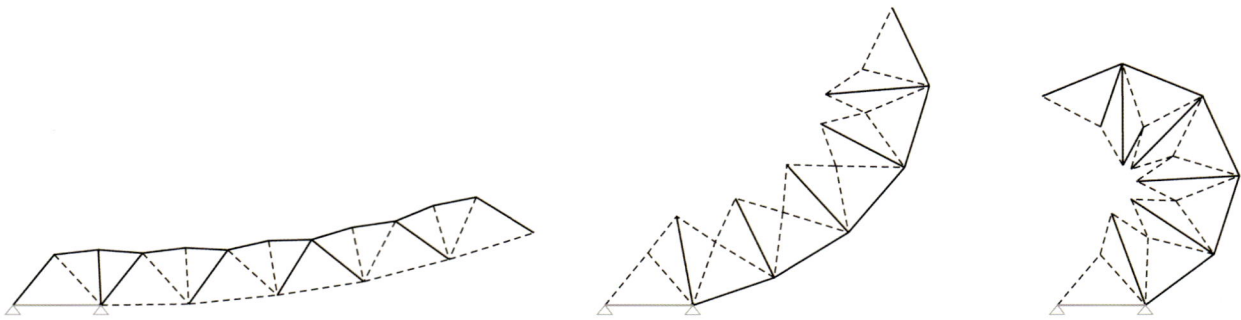

Sam Joyce, Rolling Bridge simulation, School of Architecture and Civil Engineering, University of Bath, 2008
The program simulates the physical behaviour of the Rolling Bridge, comparing the abstracted physics of the spring system with the actual performance in the bridge. Dashed lines indicate springs in tension, while solid lines indicate springs acting in compression.

computational rules of the system.[3] What emerges is an understanding of how particles and springs interact in the context of gravity and movement. What this poses for a design paradigm is that spring systems can be very lightweight computationally allowing for configurations to be developed, varied and reconfigured while structural forces are solved instantaneously.

Mining
Considering the system in a more direct application to architecture, the questions of material and assembly arise. The spring, once relaxed into a state of force equilibrium, has implications for the type of material that it may represent in the constructed form. Being in tension or compression has obvious influences on material specification. The fate-map planning of the system provides the opportunity to link between initial setup, materiality and fabrication. In the Cylindrical Net Morphologies installation at the Architectural Association, the method of inscribing a logical identification system to the particles in the spring system allowed for ease in fabrication. Retracking through the network meant reaccessing the identification system through different looping protocols. The iterative for-loop mechanism was the fundamental programming method to accomplish this degree of control and production.

Soda, CarefulSlug structure, Sodaplay.com, 2002
Model created with Sodaconstructor, a Web-based spring modeller generated in Java. The application simulates highly modifiable 2-D tension and compression structures within an environment governed by gravity and friction. The interactive application allows for springs to be edited, added and subtracted while the structural solution is realised immediately.

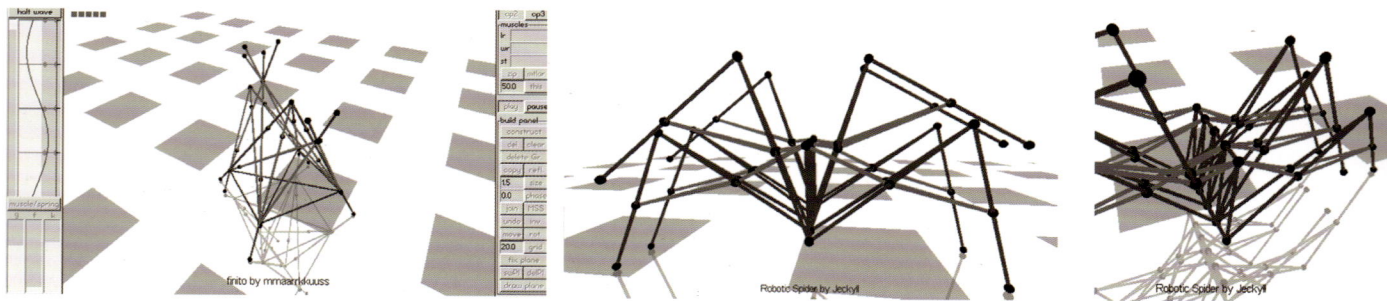

morse, Various structures in SW3d, 2008
Inspired by Sodaconstructor, SW3d.net, created by Marcello Falco in Processing, expands the functionality into three dimensions creating an environment where kinetic structures can be interactively generated and edited. Red members describe elements in tension while green members are stiff compression elements.

a

	0	1	2	3	4	5	6	7
4	(4)(0)	(4)(1)	(4)(2)	(4)(3)	(4)(4)	(4)(5)	(4)(6)	(4)(7)
3	(3)(0)	(3)(1)	(3)(2)	(3)(3)	(3)(4)	(3)(5)	(3)(6)	(3)(7)
2	(2)(0)	(2)(1)	(2)(2)	(2)(3)	(2)(4)	(2)(5)	(2)(6)	(2)(7)
1	(1)(0)	(1)(1)	(1)(2)	(1)(3)	(1)(4)	(1)(5)	(1)(6)	(1)(7)
0	(0)(0)	(0)(1)	(0)(2)	(0)(3)	(0)(4)	(0)(5)	(0)(6)	(0)(7)

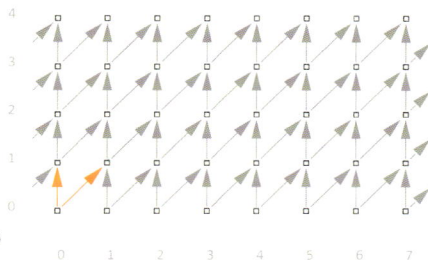

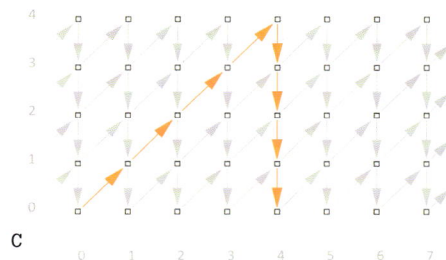

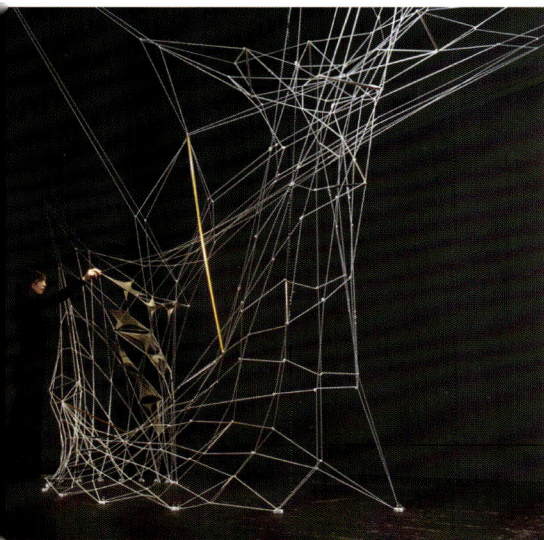

b

c

Network Mapping. Diagram (a) describes the initial mapping of particles and node identification. Diagrams (b) and (c) show the network topology of the springs and highlight the varied movement through the particle mapping, first for the connections between springs (b), and then for generating the continuous lines (strings) for fabrication (c).

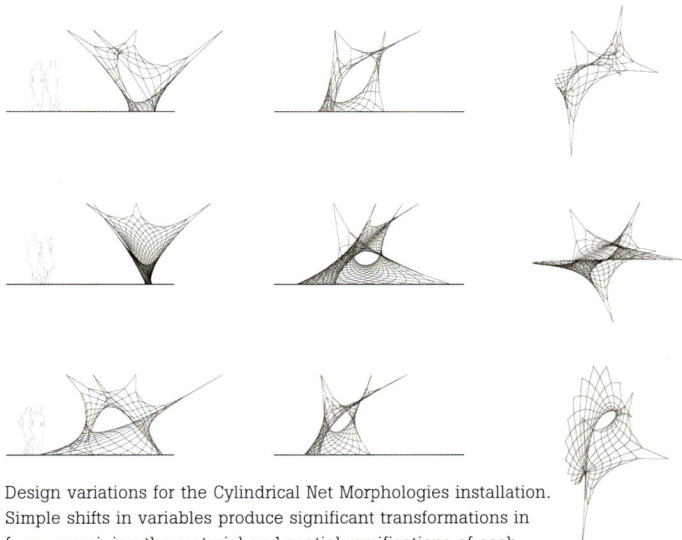

morse, Cylindrical Net Morphologies, Architectural Association, London, 2008
The tensioned net, compression elements and membrane components for this installation were developed through a computational spring system in Processing. A cylindrical network topology of springs was transformed to produce the complex physical arrangement.

Design variations for the Cylindrical Net Morphologies installation. Simple shifts in variables produce significant transformations in form, examining the material and spatial ramifications of each change. The spring system and fabrication logic are constants proving each iteration to be valid for structure and construction.

Simulation is typically seen as an engineer's tool for analysing the structural phenomena in particular material arrangements. This removes it from the realm of design investigation. When material and structural properties can be embedded into simple components of an accessible digital process, then iterative design experimentation can be engaged. An example of this has been discovered in the use of springs within the Processing programming environment. Once structural behaviour and manufacturing logics have been invested into the process, the system is able to pursue the relationship between material arrangement and its effects on space, organisation and environment. Behaviour can be expanded beyond technical and engineering realms encompassing particularities of architectural space. ∆+

Notes

1. Simon Greenwold, 'PSystem: Particle Systems Plugin for Processing', Lecture at the Form-Finding and Structural Optimization: Gaudi Workshop, MIT, Cambridge, MA, 2004.

2. Helmut Pottman et al, *Architectural Geometry*, Bentley Institute Press (Exton, PA), 2007.
3. Casey Reas and Ben Fry, *Processing: A Programming Handbook for Visual Designers and Artists*, MIT Press (Cambridge, MA), 2007.

Sean Ahlquist and Moritz Fleischmann formed morse, a design group aimed at linking research, innovation and professional practice through new design strategies, in 2008. Their work was most recently exhibited at the Architectural Biennale Beijing (2008). Both are graduates of the Emergent Technologies and Design Programme at the AA in London, and are currently PhD candidates at the Institute for Computational Design (ICD) at the University of Stuttgart, researching computational strategies for material systems and spatial configurations. Ahlquist previously founded the firm Proces2 in San Francisco and taught at UC Berkeley and California College of the Arts. He has collaborated on projects such as Airspace Tokyo with Thom Faulders, and the Jellyfish House with IwamotoScott Architecture. Ahlquist's and Fleischmann's common interest in digital tools, techniques and processes has led to collaborations with design offices around the world, among them OMA and RTKL.

'Unit Factor' is edited by Michael Weinstock, who is Academic Head and Master of Technical Studies at the Architectural Association School of Architecture in London. He is co-guest-editor with Michael Hensel and Achim Menges of the *Emergence: Morphogenetic Design Strategies* (May 2004) and *Techniques and Technologies in Morphogenetic Design* (March 2006) issues of *Architectural Design*. He is currently writing a book on the architecture of emergence for John Wiley & Sons Ltd.

Complex Systems Architecture

Neil Spiller reveals the complexity of architectural systems to us, or is it the complexity of the architectural mind? For Spiller, the potential of a single notation as a seed for a truly ecological architecture provides an essential catalyst, triggering a multifaceted musing that takes in diagrams, Paul Preissner's new competition entry for the Taiwan Centre of Disease Control and his future work with Dr Rachel Armstrong on complex biological systems.

In recent weeks my mind has been focused, by synchronistic events, on the potential of architectural design notation to adequately describe existing sites and networks – a notation that far exceeds the primitive methodologies of traditional site analysis that have been taught in architectural schools since their inception. So might it be possible to use the same notation to proactively design a 'pre-architecture' that then becomes a truly ecological and sustainable architecture?

Simultaneously, again, I've been noting my rather wacky 'Communicating Vessels' for Mark Garcia's forthcoming book on architectural diagrams; concerning myself with complex dynamics, changing fields and sensitivities of objects and sites over time, symbolism and open-ended architectural systems.

Some of these ideas resurfaced when viewing my friend, Chicago-based architect Paul Preissner's new competition entry for the Taiwan Centre of Disease Control for a site within the Hsinchu Biomedical Science Park. In the competition proposal (2008), he describes the proposal thus:

Our proposed design for the Taiwan CDC creates a new way of organizing complex space to enable advanced research to be conducted in the most advantageous of ways. Through developing each laboratory as an individual project block connected into a continuous

Paul Preissner Architects, Centre of Disease Control, Taiwan, 2008
Roof plan, elevation and aerial perspective. Paul Preissner and his team utilise parametrics to develop architectural forms that are characterised by sleek skins or perforated faceted forms that use computer manufacturing and ideas of hybridisation of space, programme and the virtual and the vital to facilitate flows of users, technology and space.

Neil Spiller, Communicating Vessels Project Notation, Fordwich, Kent, 2008

Augmented reality drawing showing vista sculpture and the augmented reality field that senses the sculpture's position in relation to other moving ecologies within the garden, and forces the sculpture to avoid them.

Overall vista diagram showing a particular vista and its time-based dynamics with a project that harvests space-time vectors and grows an aerogel landscape.

volume of development, each program aspect is able to maintain its identifiable autonomy, while still benefiting from close proximal relationships to other aspects of the research centre. Through enabling a new openness of communication, the CDC will operate in a much higher paced experimental research lab. Furthermore, this places a new importance on the notion of communication and democratic forms of organization to create a lab tasked with understanding and solution finding for some of the most critical issues in contemporary public health. The project cascades down the site, providing a seamless plaza area to connect the project into the centre of the research park that it is a part of. The building mass lifts and bends to allow for open movement at grade, and provides outdoor courtyards throughout the project acting as researcher lounges.

Preissner highlights, in a sense, the architectural problem of our age, that of encouraging information sharing, interdisciplinary speculation and the synthesis of the arts and science – a true Third Culture.

One is also reminded of my hero Cedric Price's thought that once an institution builds a headquarters it is the end of its period of innovation. Recent problems with banks come to mind.

Simultaneously again, I've been collaborating with the amazingly proactive ambassador of the Third Culture, Dr Rachel Armstrong, science-fiction author, sometime medical consultant to performance artists Stelarc and Orlan, and occasional guest-editor of *AD*. We are developing architectural research within my AVATAR research group at the Bartlett that concerns itself with complex biological systems. We will be engineering cells near you soon, but more of that later. In conversation with biological and medical colleagues at University College London it has become apparent that architects are uniquely qualified to design these fleeting coagulations of cross-disciplinarity that exist in time both virtually and vitally.

How might we do this? Some of the researchers in biological complexity, evolution and organism robustness suggest that the biological world (and I would contend the room, the street and the city) could be seen as a series of interlinked 'hubs' that have varying connectability over time and between one another, an ecology of potential buffeted by environmental factors that develops organisms or structures that resist or assimilate this interference. This concept might be our precedent model.

We are at another of the important perturbations in technology and epistemology that seems to affect us so often these days. Cell biology is the new cyberspace and nanotechnology. Once we fully understand the exact nature of how our world makes us and, indeed, how it sometimes kills us, we will be able to make true architectures of ecological connectability.

This is our profession's future. Small steps have been made, but much more remains to be done. *Δ*+

Neil Spiller is Professor of Architecture and Digital Theory and Vice Dean at the Bartlett, University College London. His article 'Spatial notation and the magical operations of collage in the post-digital age' is to be published in Mark Garcia (ed), *Diagrams of Architecture*, John Wiley & Sons Ltd, May 2009.

McLean's Nuggets

Not Pedestrian

Eschewing its pedestrian clients, the pavement may be about to transform itself into a more multipurpose substrate. A recent installation by Dr Beau Lotto (University College London) and plastics innovator Kees van der Graaf sees specially fabricated photovoltaic paving stones utilising local waste glass products to create an energy-producing paving system in east London. The Beacon project,[1] commissioned by the Shoreditch Trust, is in part an experiment to see 'whether it is possible to turn the very ground we walk on into sites for generating energy'.[2] It also forms part of neuroscientist Dr Lotto's explorations into how we perceive and understand our visual environment. Meanwhile, in the Netherlands, researchers at the University of Twente have produced a prototype road paving system that purifies (or cleans) the air.[3] Based on an original project by the Mitsubishi Industrial group of Japan,[4] modified concrete blocks are fabricated with an outer layer of the photo-catalytic

Detail of pavement-based photovoltaic panels. The Beacon project, Shoreditch, London, 2008.

ingredient titanium dioxide, which in sunlight will metabolise harmful nitrogen oxides contained in vehicle exhausts into more benign nitrates. In a controlled test, a street in Hengelo will be laid with 50 per cent standard block pavers and 50 per cent air-purifying blocks, and then be physically tested and compared for effectiveness of particulate cleansing. Incidentally, titanium dioxide is widely used in white paint, and as an ingredient of toothpaste, sunscreen and food colouring. In another street, cleansing/harvesting opportunity research student Angela Murray (Department of Chemical Engineering, University of Birmingham) has highlighted the high levels of platinum and other precious metals such as palladium and rhodium contained in road dust. This high-value detritus is expelled from vehicular catalytic converters and Murray is working on a project to recover them: 'What people don't realise is that there are now almost the same levels of these metals present on our city streets as there are in the places where they are originally mined. Our streets really are paved with platinum.'[5]

Move On Up

In a socially dynamic experiment by Daniel Hirschmann and his collaborators Andy Cameron, Federico Urdaneta, Carlo Zoratti and Hans Raber, the temporary installation *Tuned Stairs*[6] at the Centre Pompidou in 2006 saw the treads of a steel stairway transformed into a kind of oversized distributed glockenspiel. With the simple addition of pressure-pad sensors, a series of ascending/descending one-note musical instruments were played by small physical actuators (solenoids), choreographed by the movement of the building's visitors and controlled by the ethical electronics of the Arduino microprocessor. Video footage of the piece shows the behaviour-altering properties of enhanced (or tuned) acoustic feedback; it also showed how the playful visitor and more object-orientated individual (who needs to get to the top) could coalesce on a single vertical transportation device. Hirschmann is currently working with interactive designer/artist Jason Bruges who, at a recent event in London, talked about the 'dwell time' or the modification of human actions that results from some of his visually dynamic public projects, concluding that 'there is money tied up in the movement and behaviour of people'.[7]

Tuned Stairs prototype using cowbells and Solenoid sounders. Fabrica exhibition (with Daniel Hirschmann), Centre Pompidou, 2006.

Visual Acuity

If then we represent our Earth as a little ball of one inch diameter then the Sun would be a big globe nine foot across and 323 yards (295m) away, that is about a fifth of a mile, four or five minutes walking.

– Cedric Price after HG Wells[8]

Price is adapting an analogy made by Wells to describe the relative scale of two differently sized objects at a not inconsiderable distance apart (152 million kilometres/94 million miles). This ability or need to make visible the scalar relations between objects built or unbuilt is the readily available tool of the designer and explains why the rubber-stamped image of a double-decker bus appears on a number of Price's drawings as a useful 'scale' of reference. The resolving power of the eye or its capacity to make out detail at distance can be described as visual acuity. The proto-scientist and polymath Robert Hooke estimated at the end of the 17th century that the limit in visual acuity angle is one minute of arc, or one-sixtieth of one degree. That is to say: 'The visual acuity limit of one minute of arc corresponds to an object interval of one and a half feet being just resolved by the eye at a distance of one mile (1.6km), or 0.075mm at the nominal near point distance of 10 inches (254mm).'[9]

The 4.2-metre (14-feet) gunmetal minute hand of the 7-metre (23-foot) diameter clock face of London's Big Ben is approximately 304 millimetres (1-foot) wide. This kind of information

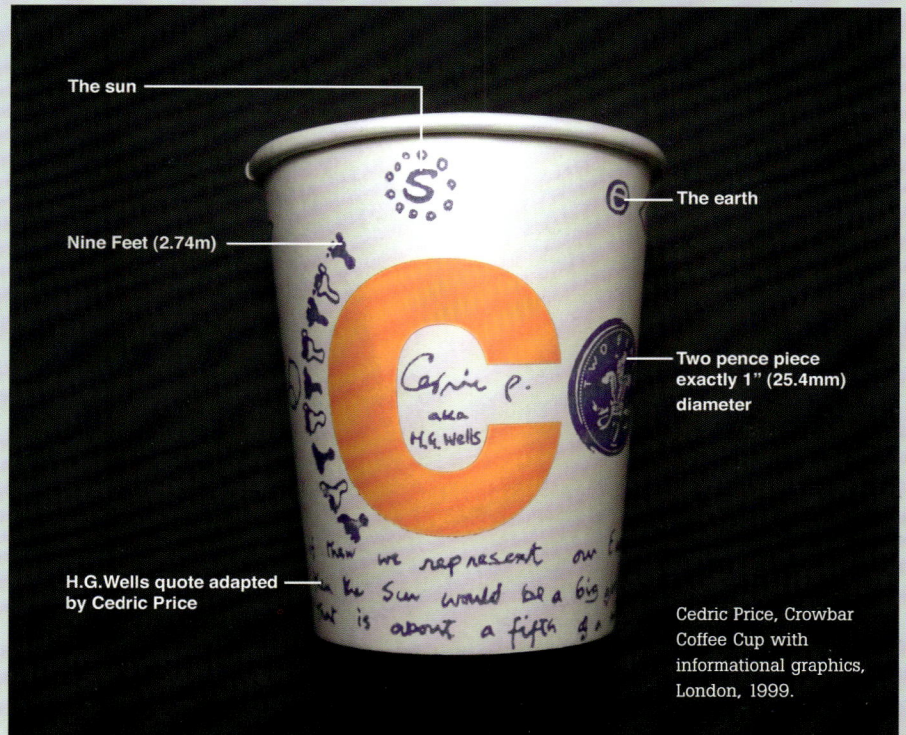

The sun

Nine Feet (2.74m)

H.G.Wells quote adapted by Cedric Price

The earth

Two pence piece exactly 1" (25.4mm) diameter

Cedric Price, Crowbar Coffee Cup with informational graphics, London, 1999.

used as a design tool may be useful when designing at distance or designing an object to be viewed from a distance. When reviewing an architectural drawing (an elevation even), if such things need still exist, are you looking at your creation from the conceptual bus stop over the road or at an un-focusable near proximity? Or to rephrase: why buy that bigger television or computer screen when you could just sit much closer to the diminutive one you already own? The success of the small viewed close can be seen with any hand-held computer/communication device (phone). Δ+

'McLean's Nuggets' is an ongoing technical series inspired by Will McLean and Samantha Hardingham's enthusiasm for back issues of *AD*, as explicitly explored in Hardingham's *AD* issue *The 1970s is Here and Now* (March/April 2005).

Will McLean is joint coordinator of technical studies (with Pete Silver) in the Department of Architecture at the University of Westminster. He recently co-authored, also with Pete Silver, the book *Introduction to Architectural Technology* (Laurence King, 2008).

Notes

1. http://www.lottolab.org.
2. Ibid.
3. See 'Air-Purifying Paving', *The Engineer*, 13 August 2008.
4. See 'Japan Paves the Way to Clean Air', *The Engineer*, 15 January 2000.
5. 'Regional Roundup', *Environment Times*, Vol 12, Issue 4, 2008, p 9. See www.environmenttimes.co.uk.
6. http://www.plankman.com/projects/tuned_stairs/.
7. Jason Bruges, 'Building Ephemeral Cities', Lecture at the University of Westminster School of Architecture and the Built Environment, 23 October 2008.
8. From Samantha Hardingham (ed), *Cedric Price: Opera*, Wiley-Academy (London), 2003, p 110.
9. BK Johnson, *Optics and Optical Instruments*, Dover Publications (New York), 1960, p 43.

The Sustainable Disco

Valentina Croci describes how Dutch artist Daan Roosegaarde's Sustainable Dance Floor (SDF) provides the centrepiece for an eco-club in Rotterdam that builds low energy consumption into the attraction. The square dance floor uses 'mechatronic' sensors to convert kinetic movement into electricity when people dance. Movement is not only part of the thrill, but is literally electrifying.

Kossmann.dejong and Döll-Atelier voor Bouwkunst, Club Watt, Rotterdam, The Netherlands, 2008
View of the performance and concert space, located on the ground floor. The space is organised on multiple levels, with bars and an open area below the stage.

The 30-square-metres (323-foot) Sustainable Dance Floor (SDF) is located in the basement. The project was designed by Daan Roosegaarde of Studio Roosegaarde. When dancers produce 20 watts, the floor powers itself. The electricity generated by the dancers is displayed on the 'Energy Meter' at the edge of the dance floor.

Club WATT, which boasts the highest percentage of energy savings for a place of this size, opened in Rotterdam in September 2008. The design of the club took almost two years and was managed by the Dutch company Sustainable Dance Club (SDC), a group that offers consultancy and design services for entertainment spaces. Club WATT can accommodate 2,000 people on two different levels and includes the 150-seat Teaterzaal theatre, the Lulu café-restaurant, a garden in the interior courtyard, the dance club in the basement and a roof garden. The interiors were designed by the Dutch architecture offices Kossmann.dejong and Döll-Atelier voor Bouwkunst.

The club is lit by LED lighting fixtures that use up to 85 per cent less energy than standard. The project also features a system to recycle rainwater for use in the bar and washrooms (the urinals in the men's washroom are waterless) and a device that controls CO_2 emissions. Altogether this allows for an estimated 30 per cent reduction in the total energy consumed by the club – 50 per cent less water and 30 per cent less CO_2 – compared with a traditional club of the same size. What is more, the club has its own policies for recycling, reducing consumption of paper and packaging by, for example, serving its drinks in 'ecocups'. Staff and catering suppliers are also selected in light of the sustainable approach adopted by the club. The cost of waste produced in a year is thus expected to drop by 50 per cent. In analogous terms, the heating and air-conditioning system is centralised and features a heat-recovery unit. In the future it is hoped that the use of fossil fuels can be reduced in favour of the use of alternative energies, for example solar, wind and biomass.

One of the most interesting innovations at Club WATT is the Sustainable Dance Floor (SDF) developed by the Dutch artist Daan Roosegaarde in collaboration with the TU in Delft, the TU in Eindhoven, Paulides BV and Advanced Electromagnetics BV. The 30-square metre (323-foot) floor is made of floating 65 x 65 centimetre (25.6 x 25.6 inch) tiles that convert kinetic movement into electricity when people dance. Movement is captured by 'mechatronic' sensors (a combination of mechanical and electronic engineering) and converted into electricity by a dynamo. The more a dancer moves, the more electricity is generated, up to a maximum of 20 watts – at which point the energy provided is sufficient for the dance floor to power itself.

Roosegaarde's studio worked on the interface and the experience of the user. The aim was to avoid screens with writing on them, and instead to emphasise the act of movement and the fact that we can produce energy simply by dancing. Each tile is backlit with LED point lights. Points, instead of lines, illuminated only 10 times per second, allow the human retina to perceive fields of continuous colour, while simultaneously consuming less energy. The LEDs are connected to a

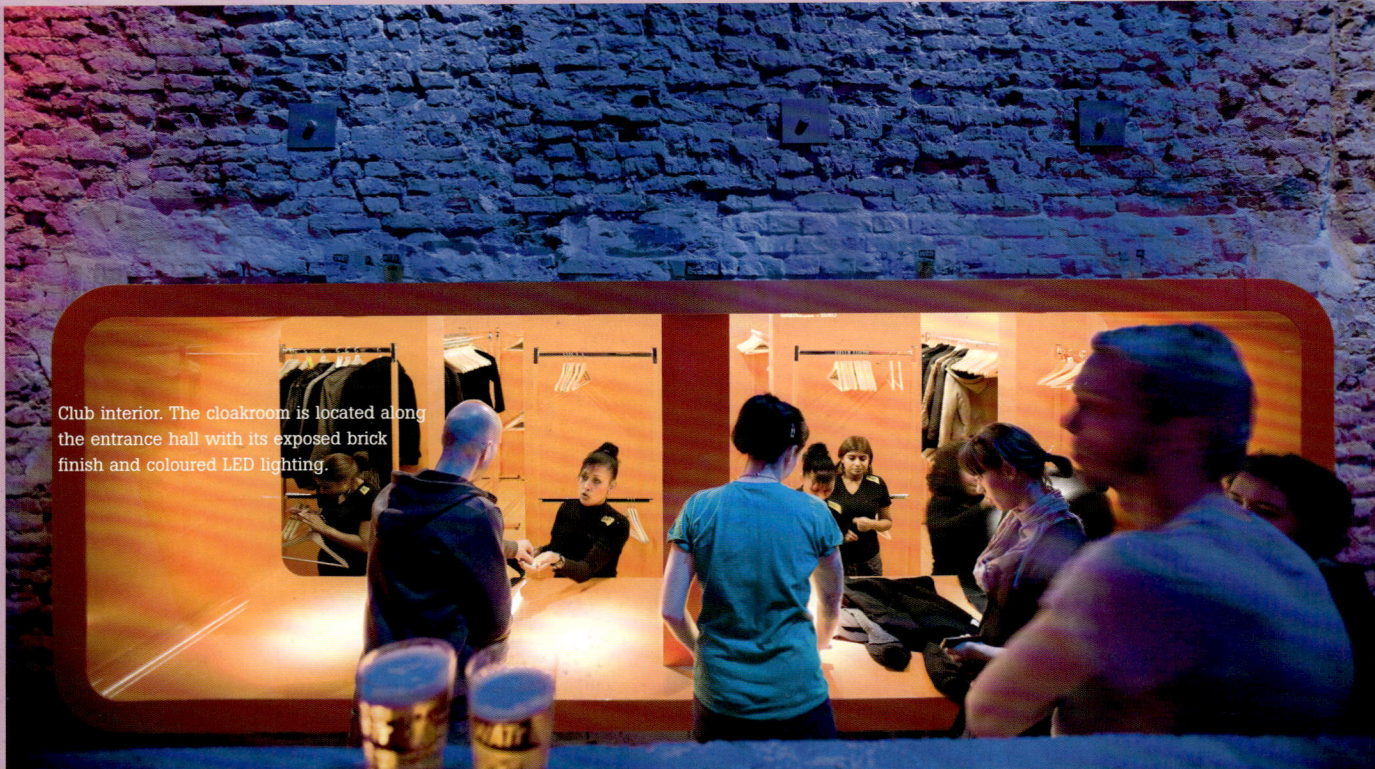

Club interior. The cloakroom is located along the entrance hall with its exposed brick finish and coloured LED lighting.

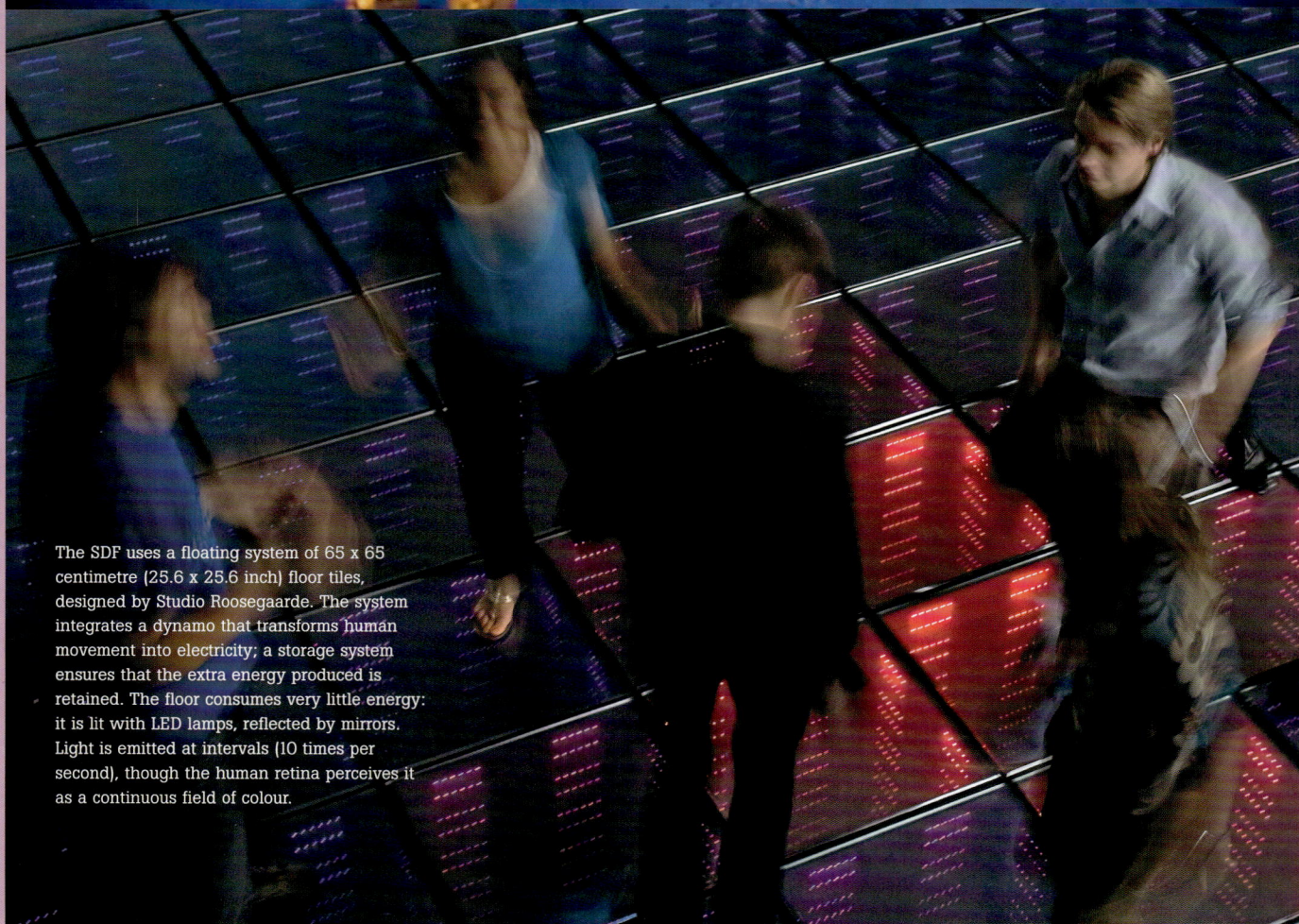

The SDF uses a floating system of 65 x 65 centimetre (25.6 x 25.6 inch) floor tiles, designed by Studio Roosegaarde. The system integrates a dynamo that transforms human movement into electricity; a storage system ensures that the extra energy produced is retained. The floor consumes very little energy: it is lit with LED lamps, reflected by mirrors. Light is emitted at intervals (10 times per second), though the human retina perceives it as a continuous field of colour.

Detail of the cash desk, a freestanding glass cube.

The central corridor provides access to the restaurant, the theatre, the internal garden and the dance club in the basement. The roof garden also overlooks the corridor.

grid of electronic microchips that increase the intensity of the light in proportion to an increase in movement. In parallel to this, 'warning lights' (small green dots) appear and move in the specific area in which energy is being produced, which is measured by and displayed on an 'Energy Meter' connected to the floor.

The interface is what defines the user's experience. Roosegaarde says: 'I concentrated on the social meanings and rituals related to this type of environment, focusing on the idea of kinematics: to activate the floor's dynamic lighting system you have to dance; the more you dance, the more profound the experience becomes, something that is demonstrated by the flashing floor tiles. We worked with a different notion of sustainability: not so much a forced rule of behaviour, or the reduction or recycling of something that already exists, as much as an activity of creating energy connected with an entertaining experience.' The interaction between the floor-management software and the user creates a sort of play between dancers, inciting them to move. Sustainability thus becomes a mutual action, shared by a group of people who co-control the system and co-produce energy. The Energy Meter adds the final touch to users' awareness of what is taking place: a normal action is loaded with a more significant meaning – the idea of sustainability – which is now a social imperative.

'Connecting interactivity with sustainability is a significant challenge,' Roosegaarde adds. 'SDF is both a product with a function and an object of art that reflects my personal research. In my opinion, the interaction between man and machine is not a question of technology and its ability to react to the actions of the user – this is a reflexive process – but rather a tool for questioning our behaviour. The interactive installation, above all if it is located in a public place, introduces a sensation of "displacement" and a new dialogue between the user and

his or her physical context. The ways in which we relate to the interactive object depends on where we find it and the cultural system of reference. For example, another of my works, entitled "Dune", was witness to widely varying attitudes from different users when placed in different geographical contexts. This conceptual complexity can be transferred to everyday objects, such as the SDF.'

The SDF and, more generally, the activities of SDC are, in fact, an open programme. SDC proposes a general design philosophy and offers know-how and technical consultancy. Their focus is not on offering serial versions of clubs, but projects that match their specific context of reference. What is more, for each club they will work with a different architect to define the interior detailing. In analogous terms, each new SDF will feature original aesthetic and technical solutions. SDF is not a finished product, but ongoing research that concentrates on capitalising on the behaviour of its users to create energy. 'SDC's idea is still in the embryonic phase, and can be increased: for example, using the energy accumulated in the mechanical and lighting systems in each club. We are studying ways to take advantage of the kinematics present in other parts of the club, and how to capture the energy produced by the shakers used by bar staff,' Roosegaarde continues.

For now, Club WATT remains a costly project, with expenses to be recovered over four years. However, LED technology, like the other sectors of electronics and 'clean' energy technologies (solar and wind power) are fast becoming more accessible. This makes room for sustainability in architectural and interaction design. For Roosegaarde the SDF project has opened the door to unexplored territory: sustainability as a theme of artistic research. ⚙+

Translated from the Italian version into English by Paul David Blackmore

Valentina Croci is a freelance journalist of industrial design and architecture. She graduated from Venice University of Architecture (IUAV), and attained an MSc in architectural history from the Bartlett School of Architecture, London. She achieved a PhD in industrial design sciences at the IUAV with a theoretical thesis on wearable digital technologies.

Replacing a Beloved Building with a Hybrid
Paresky Student Center, Williams College

Jamie Horwitz describes how Polshek Partnership's design for a new student centre at Williams College, a small liberal arts college in Williamstown, Massachusetts, has successfully integrated 'environmentally sensitive practices into the choreography of life'. Catering not only to students' 24-hour needs, its dining facilities tap into a network of local food producers within its far-reaching sustainable programme.

Architecture's role in social memory is well illustrated by Williams College alumni's attachment to Baxter Hall. A live webcam permitted this dispersed community to observe the building's demolition via the college homepage.

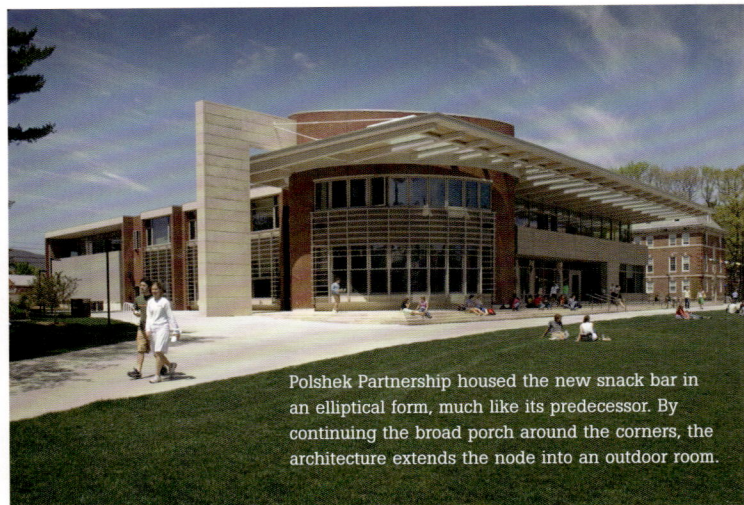

Polshek Partnership housed the new snack bar in an elliptical form, much like its predecessor. By continuing the broad porch around the corners, the architecture extends the node into an outdoor room.

The students were skeptical that a brand-new building could ever have soul, but from the moment Paresky opened its doors, it became an important gathering place for the campus community, a site of both energy and ease.

Darra Goldstein, Francis Christopher Oakley Third Century Professor of Russian, Williams College, 2008[1]

Campus design is notoriously challenging. The shifting interests of students, alumni, donors, operations, dining services, administrators, faculty, neighbourhood review boards and local activists have had a chilling impact on the most notable design firms. In the case of the recently completed Paresky Student Center, all parties agree: Polshek Partnership never tired of the spirit of democracy at Williams College.

The challenge at Williams was to design a new structure loaded with programme – kitchen and dining service facilities, seating in three venues, a performance space for lectures and stage production – in the middle of a tight cluster of landmark buildings. Expanding and upgrading the campus centre for 2,000 students at this selective liberal arts college in the hills of western

Massachusetts involved producing a building that could exert the same centripetal force as its predecessor, the 'beloved' Baxter Hall, and particularly its elliptical gathering spot the Snackbar. At the same time, the architects were charged with producing a building in keeping with Williams' increasingly bold institution-wide commitment to environmentally sustainable practices, a commitment that is among the strongest in US higher education.

Polshek Partnership's response mixes traditional building methods with the most technologically advanced systems, melding the two into existing social and spatial patterns. From its outermost edges, the Paresky Center appears permeable and inviting. While there is nothing explicitly hospitable about the 9.8-metre (32-foot) cantilevered steel-and-glass canopy that forms a double-height porch over wide gentle steps, students nevertheless drag chairs, laptops and lunch out on to this spatial threshold. Designed to extend the central campus green, the Paresky Center spills an exceptionally friendly and gracious living room on to the lawn.

An interior of similarly layered zones – rather than walls – forms interlocking horizontal and vertical niches. Partially enclosed, with adjustable comfortable seating, these spaces are perfectly suited to studying, socialising and shifting between the two. The condition is found repeatedly in a new generation of elegant campus centres such as Weiss Manfredi's at Smith College, and Rem Koolhaas' OMA design of IIT's McCormick Tribune Campus Center. Yet, while the others emphasise movement and urban passageway, Polshek's Paresky Center invites people to stay a while. It is a place to relax, not only to study and eat or chat and dash. It is a place to form relationships that make college last a lifetime.

Given the Williams College tradition to keep the student centre open 24 hours a day, Paresky requires an energy strategy that addresses drastic fluctuations in use, as well as the fierce winters and humid summers of western Massachusetts. In addition to a super-insulated shell (the 30.5-centimetre/12-inch roof has an R value of 42, and the walls provide R-21) and a solar shade built into the entire south facade, the new building is outfitted with occupancy sensors for heating, cooling and lighting to minimise energy waste, particularly at night. Daylight sensors also determine the need for artificial illumination, and carbon-dioxide sensors monitor and adjust ventilation.

A great hall and fireplace with an 8-metre (26-foot) tall Vermont slate chimney mass – the celebrated architectural feature of Paresky – also benefit from a hybrid of traditional and new forms of sustainability. By embedding the HVAC system in the floor instead of the ceiling (using honeycombed radiant floors) this 11-metre (36-foot) high timber space remains open to the surrounding dining facilities and lounges on two levels with natural light on four sides from clerestory and skylights. Unfortunately, the fireplace has yet to be lit because of an unresolved conflict with the air-handling system.

Paresky's largely successful integration of environmentally sensitive practices into the choreography of life at Williams receives its fullest expression in the design of the dining service facilities. From a loading dock that is perfectly suited for deliveries by small growers (who take away food waste for composting and convert used cooking oil into the biofuel for their trucks) to a kitchen where huge deliveries of organic produce are flash frozen in vacuum-packed bags, and pear tarts and strudels are rolled out on massive butcher blocks, we find the architecture of institutional dining contributing to fine cuisine and landscape renewal, as well as regional economic development. Under the guidance of Robert Volpi, the Director of Food Services at Williams College (and a recipient of a Renew America Award for national environmental stewardship), Richard Olcott of Polshek Partnership and Kathleen Seely of Ricca Newmark Consultants, the Paresky Center is designed to integrate the entire food axis. This system of procurement, delivery, storage, preparation, service, clean up and disposal brings a campus to the table at the same time as it crosses that most impermeable boundary – the one between 'town and gown', or the college and its regional economy and landscape.

Until it is mandated, environmental change in the US can only happen if it is successfully negotiated in individual buildings like Paresky. This is a model of sustainable architecture in which the physical fabric of the building is barely distinguishable from the policies and desires that shaped it. It is worth remembering that the design process is itself a unique method through which a community like Williams College – with its multiple ambitions and agendas – can coalesce into the remarkable social generator that has produced the Paresky Center. ⚏+

Jamie Horwitz is faculty adviser to the Iowa State University Solar Decathlon team and holds the 2009 Frederick Lindley Morgan Chair of Architectural Design at the University of Louisville, Kentucky. Her co-edited collection, *Eating Architecture*, first published in 2004, was republished in paperback by the MIT Press in 2006.

Note
1. Darra Goldstein, Francis Christopher Oakley Third Century Professor of Russian, Williams College, and Editor in Chief of *Gastronomica, The Journal of Food and Culture*, email correspondence with the author, October 2008.

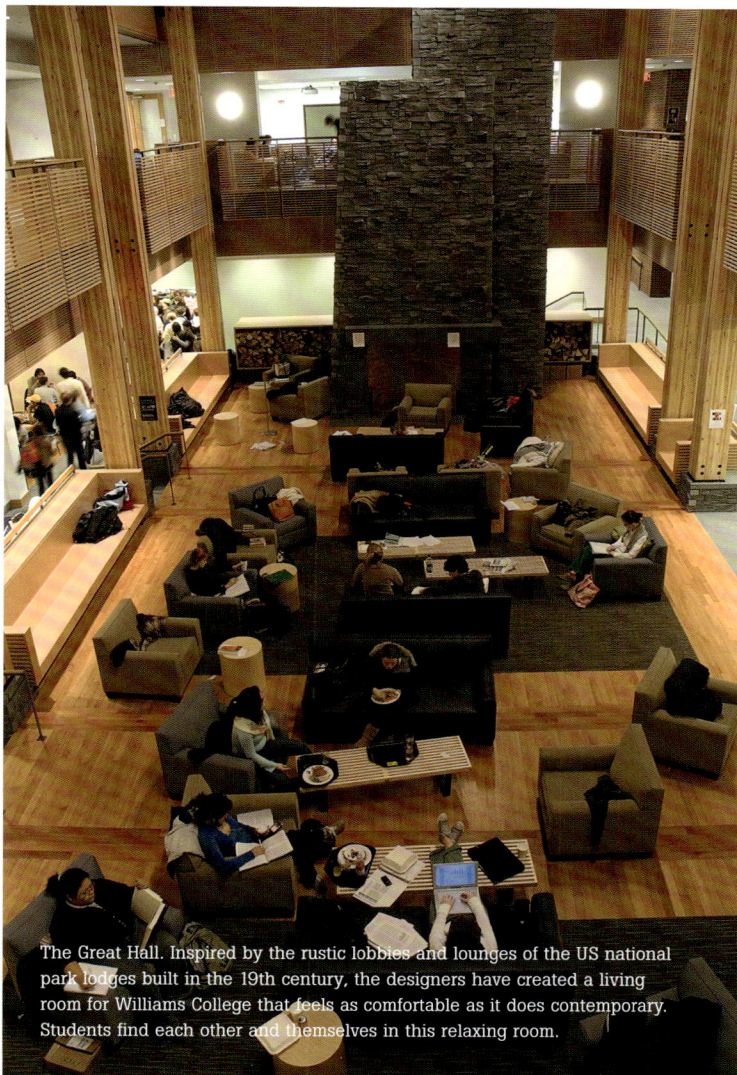

The Great Hall. Inspired by the rustic lobbies and lounges of the US national park lodges built in the 19th century, the designers have created a living room for Williams College that feels as comfortable as it does contemporary. Students find each other and themselves in this relaxing room.

Architectural Design **Theoretical Meltdown** January/February 2009

What is Architectural Design?

Launched in 1930, *Architectural Design* is an influential and prestigious architectural publication. With an almost unrivalled reputation worldwide, it is consistently at the forefront of cultural thought and design.

Architectural Design is published bimonthly. Features include:

Main section
The main section of every issue functions as a book and is guest-edited by a leading international expert in the field.

△+
The △+ magazine section at the back of every issue includes ongoing series and regular columns.

Truly international in terms of the subjects covered and its contributors, *Architectural Design*:

- focuses on cutting-edge design
- combines the currency and topicality of a newsstand journal with the rigour and production qualities of a book
- is provocative and inspirational, inspiring theoretical, creative and technological advances
- questions the outcomes of technical innovations as well as the far-reaching social, cultural and environmental challenges that present themselves today

How to Subscribe

With 6 issues a year, you can subscribe to △ (either print or online), or buy titles individually.

Subscribe today to receive 6 issues delivered direct to your door!

£198 / US$369	institutional subscription (combined print and online)
£180 / US$335	institutional subscription (print or online)
£110 / US$170	personal rate subscription (print only)
£70 / US$110	student rate subscription (print only)
To subscribe:	Tel: +44 (0) 843 828
	Email: cs-journals@wiley.com

To purchase individual titles go to:
www.wiley.com

Erratum

In the article 'Horizons of User-Centred Design' by Valentina Croci in the *AD Neoplasmatic Design* issue (Nov/Dec 2008), the 'Local Barometers' and 'Plane Tracker' projects were miscredited in the figure captions. The projects are by the Interaction Design Studio, Goldsmiths University of London (William Gaver, Andy Boucher, Andy Law, Sarah Pennington, John Bowers, Jake Beaver, Tobie Kerridge and Alex Wilkie) with Jan Humble (University of Nottingham) and Nicholas Villar (University of Lancaster). We apologise for this error to our readers and those directly involved.